Molecular Biology
Intelligence
Unit

Mobile Genetic Elements in Metazoan Parasites

Paul J. Brindley, PhD
Department of Microbiology, Immunology
and Tropical Medicine
George Washington University Medical Center
Washington, DC, USA

CRC Press
Taylor & Francis Group
Boca Raton London New York

CRC Press is an imprint of the
Taylor & Francis Group, an informa business

Mobile Genetic Elements in Metazoan Parasites

Molecular Biology Intelligence Unit

CRC Press
Taylor & Francis Group
6000 Broken Sound Parkway NW, Suite 300
Boca Raton, FL 33487-2742

First issued in hardback 2019

ISBN-13: 978-1-58706-093-9 (hbk)

Visit the Taylor & Francis Web site at
http://www.taylorandfrancis.com

and the CRC Press Web site at
http://www.crcpress.com

Library of Congress Cataloging-in-Publication Data

Mobile genetic elements in metazoan parasites / [edited by] Paul J. Brindley.
p. ; cm. -- (Molecular biology intelligence unit)
Includes bibliographical references and index.
ISBN 978-1-58706-093-9
1. Mobile genetic elements. 2. Metazoa. 3. Host-parasite relationships--Genetic aspects. 4. Trematoda. 5. Schistosoma. I. Brindley, Paul J. (Paul James), 1954- II. Series: Molecular biology intelligence unit (Unnumbered : 2003)
[DNLM: 1. Parasites--genetics. 2. DNA Transposable Elements--genetics. 3. Retroelements--genetics. QX 4 M687 2009]
QH452.3.M643 2009
572.8'6--dc22

2009029022

Dedication

To Victoria and Meredith

About the Editor...

PAUL J. BRINDLEY is a Professor of Microbiology and Tropical Medicine at the Department of Microbiology, Immunology and Tropical Medicine, George Washington University School of Medicine and Health Sciences, Washington, DC. His research interests include genomics/proteomics of helminth parasites, including endogenous mobile genetic elements, gene manipulation and transgenesis of helminth parasites, and infection related cancers. He has held academic appointments and fellowships in Australia, Ireland and the USA. He is deputy editor of *PLoS Neglected Tropical Diseases.*

CONTENTS

EDITOR

Paul J. Brindley, PhD
Department of Microbiology, Immunology and Tropical Medicine
George Washington University Medical Center
Washington, DC, USA
Email: mtmpjb@gwumc.edu
Chapters 6, 7 and 9

CONTRIBUTORS

Note: Email addresses are provided for corresponding authors of each chapter.

Young-An Bae
Department of Molecular Parasitology and Center for Molecular Medicine
Sungkyunkwan University School of Medicine
and
Samsung Biomedical Research Institute
Suwon, Korea
Chapter 3

Joanna Bridger
Laboratory of Nuclear and Genomic Health
Centre for Cell and Chromosome Biology
School of Health Sciences and Social Care
Brunel University
Uxbridge, Middlesex, UK
Chapter 4

Craig J. Coates
Department of Entomology
Texas A&M University
College Station, Texas, USA
Email: c-coates@tamu.edu
Chapter 5

Claudia S. Copeland
Department of Tropical Medicine
Tulane University Health Sciences Center
New Orleans, Louisiana, USA
and
Center for Medical, Agricultural and Veterinary Entomology, USDA
Gainesville, Florida, USA
Chapters 6, 7 and 9

Ricardo DeMarco
Departamento de Física e Informática
Instituto de Física de São Carlos
Universidade de São Paulo
São Carlos, Brazil
Email: rdemarco@ifsc.usp.br
Chapter 8

Gerardo Ferbeyre
Université de Montréal
Département de Biochimie
Montréal, Quebec, Canada
Email: g.ferbeyre@umontreal.ca
Chapter 1

Wannaporn Ittiprasert
Biomedical Research Institute
Rockville, Maryland, USA
Chapter 4

Matty Knight
Biomedical Research Institute
Rockville, Maryland, USA
Email: mknight@afbr-bri.com
Chapter 4

Yoon Kong
Department of Molecular Parasitology and Center for Molecular Medicine
Sungkyunkwan University School of Medicine
and
Samsung Biomedical Research Institute
Suwon, Korea
Email: ykong@med.skku.ac.kr
Chapter 3

Thewarach Laha
Department of Parasitology
Khon Kaen University
Khon Kaen, Thailand
Chapters 6, 7 and 9

Song Li
Department of Biochemistry
Virginia Polytechnic Institute and State University
Blacksburg, Virginia, USA
Chapter 2

Alex Loukas
Division of Infectious Diseases
Queensland Institute of Medical Research
Brisbane, Queensland, Australia
Chapter 9

Julio Masabanda
ID LABS(tm) Inc.
London, Ontario, Canada
Chapter 4

André Miller
Biomedical Research Institute
Rockville, Maryland, USA
Chapter 4

Edwin Odoemelam
Laboratory of Nuclear and Genomic Health
Centre for Cell and Chromosome Biology
School of Health Sciences and Social Care
Brunel University
Uxbridge, Middlesex, UK
Chapter 4

Nithya Raghavan
Biomedical Research Institute
Rockville, Maryland, USA
Chapter 4

Zhijian Tu
Department of Biochemistry
Virginia Polytechnic Institute and State University
Blacksburg, Virginia, USA
Email: jaketu@vt.edu
Chapter 2

Sergio Verjovski-Almeida
Departamento de Bioquímica
Instituto de Química
Universidade de São Paulo
São Paulo, Brazil
Chapter 8

PREFACE

Substantial proportions of the genome of most eukaryotes, including those of metazoan parasites, are comprised of repetitive sequences. Mobile genetic elements (MGEs) constitute much of the repetitive component, particularly the interspersed sequences. This volume of nine chapters addresses MGEs that have colonized the genomes of medically important metazoan parasites and their vectors. These MGEs are parasites of parasites. These mobile elements are compared and contrasted with those from better-studied model organisms. The authors have included plentiful, relevant, and informative illustrations and references.

Several chapters focus on mobile elements of the human schistosomes. Gerardo Ferbeyre focuses on the putative ribozyme activities on the SM alpha elements, SINE-like elements, including trans reactions catalyzed by these retroposons. Ricardo DeMarco and Sergio Verjovski-Almeida review the cut-and-paste transposons of schistosomes, including the *Merlin* and the *SmTRC1* elements, and application of schistosome DNA transposons to transgenesis and mutagenesis. Claudia Copeland, Thewarach Laha and I review the retrovirus-like retrotransposons and non-long terminal repeat transposable elements of the schistosomes, including the *Boudicca*, *fugitive* and *SR2* elements. Young-An Bae and Yoon Kong provide descriptions of retrotransposons from the human liver and lung flukes, *Clonorchis sinensis* and *Paragonimus westermani*, respectively, including the *CsRn1* long terminal repeat retrotransposon. Continuing the investigations of MGEs of parasitic worms, Thewarach Laha, Alex Loukas and I summarize current knowledge of MGEs from hookworms, including details of the *bandit* and *Transib* transposons and *dingo* retrotransposons.

Mobile elements in vectors of human parasites are also covered. Matty Knight and coauthors describe the *nimbus* retrotransposon and its discovery and other repetitive elements from the genome of the gastropod snail *Biomphalaria glabrata*, the intermediate host of *Schistosoma mansoni*. Zhijian Tu and Song Ti review MGEs of *Anopheles* mosquitoes, vectors of malaria, and other mosquitoes. They discuss practical uses of these MGEs. Craig Coates provides an expert article on transposon mediated transgenesis of mosquitoes, in particular dealing with his pioneering studies on the deployment of binary systems with the *Mos1 mariner* transposon.

MGEs from the genomes of these metazoan parasites have practical uses in transgenesis and epidemiological studies and also are valuable for annotation of the genomes of these vectors and pathogens. These issues and future directions are well covered in these articles. Lastly, I take this opportunity to thank the contributing authors, whose pioneering studies with mobile genetic elements of mosquitoes, schistosomes, hookworms, food-borne flukes and intermediate host snails have ensured this is an exciting and expanding field of endeavor. I thank them for their patience and generous contributions.

Paul J. Brindley, PhD
George Washington University
Washington, DC

Acknowledgements

My sincere thanks to my former doctoral students, especially Alec Drew, Thewarach Laha, Claudia Copeland and Maria Morales, for their interest and passion in mobile genetic elements of schistosomes and other pathogens. I also sincerely thank the Burroughs Wellcome Fund for support through a Scholar Award in Molecular Parasitology for our studies on mobile DNA.

CHAPTER 1

Transcripts from *Schistosoma* Nonautonomous Retrotransposons Possess Catalytic Activity

Gerardo Ferbeyre*

Abstract

Eukaryotic genomes are largely shaped by the presence of multiple mobile elements. A large fraction of these sequences carry a double life, alternating between DNA and RNA through the process of transcription and reverse transcription. The presence of mobile elements in eukaryotic genomes is potentially deleterious since they inactivate genes or change their expression. However, they also provide a source of variation for natural selection to act. Over the last decade we have learned that RNA molecules not only serve as messengers of the information content in DNA. RNA can have a variety of functions including the ability to catalyze chemical reactions. RNA molecules with catalytic activity are known as ribozymes and were originally found in self-splicing introns, the RNA subunit of RNase P and certain plant viroids and their satellite RNAs. The catalytic motif present in viroids was dubbed hammerhead due to the apparent shape of its secondary structure. Viroids use the hammerhead self-cleavage reaction to process the multimeric intermediates of their rolling-circle replication. Surprisingly, some repetitive elements discovered in the newt, schistosomes and crickets also encode hammerhead ribozymes. In schistosomes, the Smα family of nonautonomous retrotransposons apparently use the self-cleavage reaction to process the multimeric transcripts arising from transcription of tandem repeats. Apart from processing, this self-cleavage reaction could limit retrotransposons propagation, representing a novel mechanism of host protection against these genomic parasites. In addition, the Smα ribozymes can potentially catalyze RNA cleavage in trans as demonstrated in vitro for the mRNA coding for synaptobrevin. Other substrates predicted for the Smα ribozyme such as multimeric transcripts from Smα repeats containing mutated hammerhead motifs, the mRNA coding for a nuclear receptor like protein and the OZ retrotransposons require further study.

Retrotransposons as Mobile Elements

An important part of the genome in many eukaryotes is composed of repetitive sequences that can potentially replicate and propagate using host cell functions. Although the propagation of these sequences ultimately depends on the host, they may encode functions that contribute to their replication. Accordingly, they can be classified in three families:[1] (i) DNA transposons, (ii) autonomous retrotransposons and (iii) nonautonomous retrotransposons. A major difference between DNA transposons and retro-transposons is that the latter use an

*Gerardo Ferbeyre—Université de Montréal, Département de biochimie, E-515 C.P. 6128, Succ. Centre-Ville, Montréal, Québec H3C 3J7, Canada. Email: g.ferbeyre@umontreal.ca.

Mobile Genetic Elements in Metazoan Parasites, edited by Paul J. Brindley.

RNA intermediate in their life cycle. Retrotransposons are believed to be relics of the RNA world, a stage in the evolution of life where major biological functions were all carried out by RNA molecules.[2] Modern retrotransposons cycle between DNA and RNA stages and they must ensure the transition from DNA to RNA and from RNA back to DNA.

Retrotransposons are classified into two major classes. The LTR retrotransposons are distinguished by the long terminal repeats (LTRs) at both ends, which mediate transcription by RNA polymerase II. They also contain open reading frames coding for proteins functioning in their propagation.[1] The key enzyme for retrotransposons is the reverse transcriptase (RT) that catalyze their conversion from RNA back into DNA.[3] Reverse transcription occurs in the cytoplasm inside viral like particles and is primed by a tRNA molecule.[4] A distinctive feature of LTR retrotransposons is that they use double stranded DNA from reverse transcription to integrate into novel locations. Examples of LTR retrotransposons are the human endogenous retroviruses (HERV),[5] gypsy from Drosophila,[6] the Ty elements from *S. cerevisiae*[4] and Gulliver from *Schistosoma japonicum.*[7]

Non-LTR retrotransposons do not contain LTRs and use specific RNA polymerase II promoter signals in their 5' nontranslated region to go from DNA to a polyadenylated RNA. Since the promoter is part of the transcribed RNA, promoter signals are not lost upon retrotransposition. Non-LTR retrotransposons usually have two ORFs, one encoding a nucleic acid binding protein and one encoding an endonuclease-reverse transcriptase. Phylogenetic analysis of different reverse transcriptase sequences suggests that the ancestor of current retroelements was a non-LTR retrotransposon.[3] Reverse transcription for non-LTR retrotransposons occurs in the nucleus and requires nicking of the DNA by the endonuclease. The 3'end of the RNA transcribed from the retrotransposon displaces the nicked strand from the DNA cut by the endonuclease and base pairs with it. The 3'OH of the nicked DNA serves then as the primer for the reverse transcription. This process is known as Target Primed Reverse Transcription because it occurs on the target DNA.[8] Examples of Non-LTR retrotransposons are the LINE-1 elements of mammals,[9] the R1and R2 elements of *Drosophila*[10] and the R2Bm elements of *Bombyx mori.*[8] L1 elements are present in all mammals, and mobile elements that share structural features with L1 are found in the frog *Xenopus laevis*, bony fish, the slime mold *Dictyostelium*, maize and *Arabidopsis*. Indeed, all non-LTR retrotransposons, including L1s, are evolutionarily related to one another. This entire class of elements is at least 600 million years old and perhaps dates to the origins of eukaryotes.[11] Accordingly, L1-encoded protein domains are among the most ancient of all known functional domains. For example, the L1 endonuclease shares catalytic site residues with exonuclease III, an endonuclease found in the bacterium *Escherichia coli.*[12] Moreover, the catalytic subunit of telomerase, an enzyme that adds DNA sequences to the ends of chromosomes, is a reverse transcriptase (RT) that functions in a manner analogous to the L1 RT. A phylogenetic tree of eukaryotic RTs rooted by prokaryotic mobile elements also implies that the telomerase RT derived from retrotransposon RT.[13]

Nonautonomous retrotransposons contain promoter signals to pass from DNA to RNA but use the reverse transcriptase from other elements in their propagation.[14] In human the most abundant family of this class are the SINES (short interspersed nucleotide elements) that have a length from 90 to 300 bps and are transcribed by RNA polymerase III.[15] Like Non-LTR retrotransposons, SINES are transcribed from promoter signals that are part of the transcribed RNA. Most SINES are believed to be derived from ancestral tRNA genes that fused to the 3'-end of a LINE element.[14] A notable exception are the Alu repeats from primates which derive from 7SL RNA.[16] Hence, SINES share homologous sequences at their 3' end with LINE-1 and are believed to be propagated by the endonuclease and RT of LINE-1.[14]

Mechanisms to Restrict Retrotransposon Propagation

Transposition poses a potential threat to the host because integration into new sites in the genome can result in deleterious mutations.[9,14] In addition they generate a burden to the replication and chromosome segregation machineries. Hence, logic dictates that there should be an upper limit to the level of transposition that genomes can endure. It is therefore quite surprising the magnitude of transposable elements found in some genomes. In maize, for example, they make up more than 50% of the two billion base pairs that constitute this plant's nuclear DNA. The human genome contains an estimated 10^5 truncated and 4×10^3 full-length L1elements, which together constitute ~15% of the genome.[14] The majority of L1elements is not capable of transposition because they are truncated, rearranged or mutated. Nevertheless, abundant evidence indicates that L1 transposition continues to occur. Several examples of recent de novo transposition events have been identified largely as the result of mutations caused by the insertion of new L1 elements into functional genes.[1,9,14]

It is, of course, in the element's best interest to minimize genetic damage caused by integration, because the host's survival is necessary for its persistence. Three main mechanisms have been proposed to control retroelement propagation: (1) silencing, (2) targeted integration and (3) interference with transcription elongation.

Silencing of mobile elements can occur by blocking their transcription through promoter methylation or by RNA interference. Most of the 5-methylcytosine in mammalian DNA resides in transposons, where it inactivates promoter functions. Accordingly, it has been proposed that the main function of cytosine methylation is to suppress transcription of parasitic sequence elements.[17,18] RNA interference targets transposons by using small interfering RNAs that can mediate formation of heterochromatin,[19] mRNA destruction or inhibition of translation.[20] In *C. elegans* for example, this mechanism is particularly effective in the germ line, where it silences the propagation of the worm transposon Tc1.[21]

Targeted integration is a property of LTR retrotransposons from yeast,[22] maize[23] and perhaps many other species. The mobile elements insert tandemly in regions of the genome devoid of open reading frames. In this way they avoid insertional mutagenesis.[22] Finally, poor elongation of retrotransposon sequences can be due to specific sequence elements that block the passage of the RNA polymerase[24] or to premature polyadenylation both observed for LINE-1, retrotransposons.[25]

Functions of Retrotransposons

Two main ideas have been advanced about the functionality of retrotransposons. According to the parasitic DNA model, these sequences are able to propagate in the genomes independently of their selective value for the host. Their high numbers ensure their maintenance. Another idea suggests that despite their parasitic nature, mobile elements provide a convenient source of genomic variability for natural selection to act.[1] In addition, there is evidence for a functional role of these sequences in a variety of organisms where they have been involved in the regulation of gene expression and DNA repair.

The majority of known examples of control of gene expression by mobile elements involve the regulation of transcription. For example, L1 elements contain a sense promoter that drives the expression of the retrotransposon proteins and an antisense promoter that allows transcription of neighboring genes.[26] In the mouse, the B2 SINE family carries a polymerase III promoter for its own transcription and a polymerase II promoter that is responsible for the production of some mouse transcripts.[27] In fission yeast, components of the RNAi machinery and Clr4 histone methyltransferase repress several meiotic genes through nearby retrotransposon long terminal repeats (LTRs). The LTRs generate the RNA transcripts that trigger RNA interference against the neighboring genes.[19] Boeke and colleagues has documented that up to 79% of human genes contain L1 sequences in their introns. Since these retrotransposons block transcription elongation that may work as transcriptional attenuators of multiple genes.[24]

Evidence is accumulating that repetitive sequences play an important role in the origin of sexual chromosomes. In humans for example, the X chromosome is rich in L1 elements where they seem to play a role in X chromosome inactivation.[28,29] In addition, there are many more mobile elements in both the X and the Y chromosome than in autosomes.[30] Since there is no need for dosage compensation in the Y chromosome these sequences may have other functions in the context of the sex chromosomes.

In addition some mobile elements can regulate translation. Alu transcripts can accumulate in response to stress and stimulate translation by blocking the translational regulator PKR.[31] In addition Alu and other mammalian SINES can stimulate the translation of specific mRNAs in a PKR independent manner.[32]

Lastly, mobile elements have been suggested to play a role in DNA repair. In yeast for example, retrotransposon reverse transcriptases can mediate the repair of double strand breaks.[33] In human cells endonuclease-independent LINE-1 retrotransposition occurs at DNA lesions, resulting in retrotransposon-mediated DNA repair.[34]

Repetitive DNA and Mobile Elements in Schistosomes

Schistosomes or blood flukes are platyhelminth parasites of human and many other animals. Three members of this group, *Schistosoma mansoni, Schistosoma japonicum and Schistosoma haematobium*, infect over 200 million people worldwide. Their genome, arranged in sever pair of autosomes and a pair of sex chromosomes, contains a large proportion of repetitive sequences.[35] As many other eukaryotes, schistosomes have LTR retrotransposons, non-LTR retrotransposons and SINEs. Gulliver was the first LTR retrotransposon isolated from schistosomes. It is distantly related to the gypsy retrotransposon and seems to be present in hundreds to thousands of copies in the *S. japonicum* and perhaps also in the *S. mansoni* genome.[7] *S. mansoni* also harbors around 3,000 copies of Boudicca, another gypsy-like retrotransposon.[36] Non-LTR retrotransposons have been also isolated from Schistosomes. The SR1 elements are similar to the CR1 elements from chickens and at least 200 copies of them are interspersed through the genome of *S. mansoni.*[37] The SR2 elements from *S. mansoni*[38] and the SjR2 elements from the *S. japonicum*[39] are members of a recently described lineage of non-LTR retrotransposons typified by the RTE-1 non-LTR retrotransposon of *C. elegans.*

Nonautonomous mobile elements have also invaded Schistosomes genomes. The Smα family is present in *S. mansoni* and has tRNA homology regions. Smα seems to be arrayed in two different ways. Tandemly arrayed copies of Smα are around 330 bps in length while dispersed elements range is size from 107 to 330 bps.[40] A unique characteristic of the SMα family of SINEs is the presence of a self cleaving motif of the hammerhead type.[41] Related sequences to Smα also possessing hammerhead ribozymes motifs have been isolated from *S. haematobium, S. douthitti*[41] and *S. japonicum.*[42]

The Hammerhead Ribozyme Motif

Ribozymes are RNA molecules possessing catalytic activity. The first ribozyme was identified by Cech and colleagues as a self splicing RNA molecule present in the preribosomal RNA of *Tetrahymena.*[43] Shortly after, Altman and colleagues found one RNA enzyme that catalyzes the cleavage of tRNA precursors.[44] Studies on the replication of plant viroids, virusoids and their satellite RNAs also lead to the discovery of several self cleaving RNA motifs responsible for the processing of the multimeric products of rolling circle replication into unit-length molecules (Fig. 1A).[45] Four naturally occurring self-cleaving motifs have been identified, the hammerhead, the hairpin, the hepatitis delta virus (HDV) and the Varkud satellite.[46] In contrast to Group I introns and RNaseP RNA, which are large ribozymes, self-cleaving motifs are much smaller and range from the 40 nucleotides-long hammerhead RNA to the 160 nucleotides-long Varkud satellite RNA. All self cleaving motifs catalyze essentially the same reaction, the cleavage of a phosphodiester bond by using the 2' OH adjacent to the scissile phosphate.[47] This catalytic strategy is also used by protein-made RNases (Fig. 1B). It is thus remarkable that RNA enzymes can build active sites as efficient as protein enzymes. In fact RNA enzymes surpass proteins in this reaction. Protein RNases are generally nonspecific and can cleave multiple RNAs with little sequence specificity. Ribozymes use base pairing and tertiary folding interactions to cleave with high specificity a single phosphodiester bond.

The hammerhead ribozyme self-cleaving motif is represented in two dimensions as three helical regions radiating from a central core of unpaired nucleotides (Fig. 2). Detailed mutational

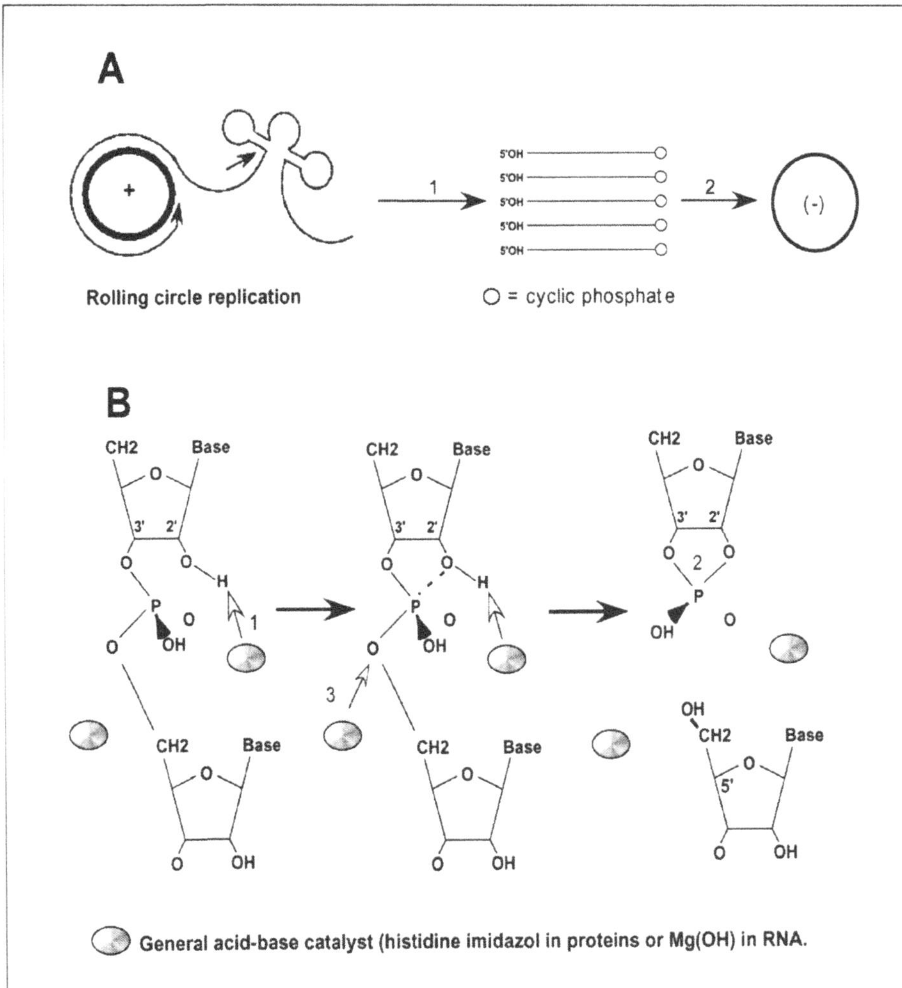

Figure 1. General properties of ribozyme self-cleaving motifs. A) Self-cleaving motifs catalyze the formation of unit-length monomers from multimeric intermediates produced after rolling circle replication. Viroids, virusoids and satellite RNAs are infectious monomer circular RNAs. The infecting strand is conventionally designed as (+) strand and it is replicated by cellular polymerases through the rolling circle mechanism. Rolling circle replication generates a multimeric linear copy of the monomer circular RNA. Self-cleaving motifs are required to convert these multimeric intermediates into monomer strands (1) which can be converted into circular RNA (-) strands by cellular RNA ligases (2) of by self-ligation also catalyzed by the RNA.[56] These (-) strands can also replicate through another cycle of rolling-circle replication-cleavage and ligation to generate the original (+) circular RNA strands.[75] B) Chemical mechanism of phosphodiester bond cleavage by self-cleaving ribozymes (proposed by Cedergren and colleagues).[47] This general phosphodiester bond cleavage mechanism is common to protein RNases and self-cleaving ribozymes. Proteins use the imidazol groups of histidine as general acid-base catalytic groups while ribozymes use hydrated magnesium ions. In either case a catalytic group (1) draws a proton from the 2'OH adjacent to the scissile phosphate bond creating an oxyanion which attacks the phosphorus atom and produces a 2'-3' cyclic phosphate (2). Simultaneously, another catalytic group (3) donates its proton to the 5'-O-CH2 leaving group.

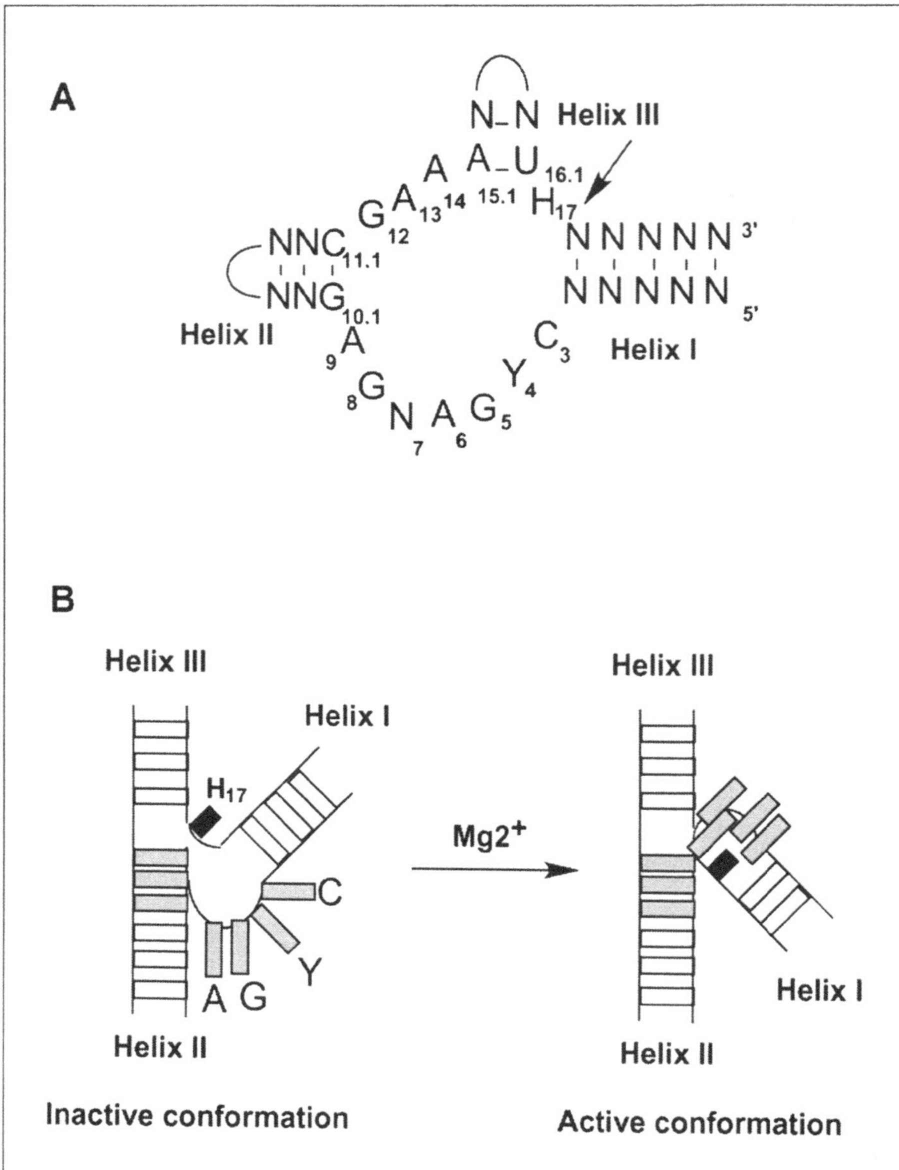

Figure 2. The hammerhead self-cleaving motif. A) Secondary structure and number system. Nucleotides require for self-cleavage activity are identified, Y= pyrimidines and H= A, C or U. Cleavage occurs after H17 and is indicated by an arrow. Most base pairs in helix I, II and III can accept any nucleotide (N). The length of the helices is variable but a minimum of base pairs is required to stabilize the structure. B) Folding intermediates and three-dimensional structure of the hammerhead self-cleaving motif according to Hamman and Lilley.[54] Magnesium ions drive a conformational change from the inactive conformer to the active one. The catalytic center of the hammerhead is formed by C3Y4G5A6, whose structure in the active conformation is similar to the uridine turn of tRNA.[52] Note how the cleavage site is away from the catalytic center in the inactive conformation.

analysis within this core revealed that most nucleotides at unpaired positions are essential for the self-cleaving reaction. Position number 7, however, may be changed to any other nucleotide[48] and position 4 can be either U (most cases) or C.[49] Position 17 at the cleavage site accepts A or C. However, U is sometimes tolerated depending on the context.[50]The ide ntity of the nucleotides in the helical regions can be changed as long as base pairing is conserved. The exceptions are the G at position 10.1 that pairs with C at 11.1, and the A at 15.1 that forms a base pair with U at 16.1.[48] Helix II can be entirely replaced by a four nucleotide connector without major suppression of the catalytic activity in vitro.[51] The crystal structures of the hammerhead ribozyme show an interaction between helix I and II,[52] so it is difficult to explain how deletion of the entire helix II is compatible with ribozyme activity. Hendry and McCall argued that the crystal structure represents an inactive conformation, suggesting that the observed interaction between helix I and II is actually blocking catalysis. They found that shortening helix I to 5 bps improved ribozyme activity in vitro and proposed that it does so because it may destabilize the inactive conformation.[53]

In three dimensions, the hammerhead is best represented as a three-way junction. The essential nucleotides are placed at the junction while the helical segments play a structural role. In branched nucleic acids, the helical segments tend to form coaxial stacking. The X-ray structure of the hammerhead showed coaxial stacking of helices II and III, with helix I adjacent to helix II, giving an overall Y shaped structure (Fig. 2B).[54] Hammand and Lilley have proposed that magnesium ions control the overall folding of the hammerhead domain between two possible conformers. At low Mg concentrations, helices I and III are adjacent and the structure is inactive. At higher Mg concentration, helix I moves closer to helix II and away from helix III (Fig. 2B).[54] Of note, in three dimensions the helix II is actually extended by non Watson Crick base pairs between nucleotides G12:A9, A13:G8 and A14:N7 at the catalytic core. This may explain why the deletion of the entire helix II does not prevent the overall folding and activity of the hammerhead ribozyme, at least in vitro. Some natural hammerheads incorporate an extra nucleotide (A or U) after A9,[55] without a significant loss in the catalytic activity. In the three-dimensional structure this extra nucleotide may bulge out of the coaxial stacking of helices II and III and should not affect the overall configuration.

Although the hammerhead is best known for its cleavage activity, it has been reported that a crosslink between helices I and II substantially increased the rate of the reverse reaction, that is self-ligation.[56] This suggests that interactions between helices I and II control a switch between cleavage and ligation in the hammerhead motif.

The Trans Reaction

Most studies to determine the minimal requirements for hammerhead mediate RNA cleavage have been done in vitro at high magnesium concentrations and using a trans format. In the trans reaction, a part of the hammerhead motif containing most of the essential nucleotides is used as enzyme. The substrate contains only the cleavage site H17 and the essential U at position 16.1 (Fig. 2).[47] In this way, the ribozyme appears to have two recognition arms that base pair with the substrate to form helices I and III. Since the only absolute requirements for cleavage in the substrate part are U16.1 and H17, it is very easy to design hammerhead ribozymes against any cellular RNA. This idea has been extensively used, with some success, for both research and therapeutics.[57] Most applications that use the hammerhead and other ribozymes have been recently replaced by RNA interference.[58] However, RNA interference has its own problems (off target effects) and the therapeutic applications of both technologies are limited by the lack of an efficient delivery system. Finding natural hammerhead motifs able to act in trans should help to identify additional RNA elements that may help to design ribozymes able to cleave cellular RNAs.

For example, in all hammerhead formulations used so far to cleave cellular mRNAs in trans, the recognition arms are designed to stay unpaired. However, natural hammerhead ribozymes do not have single stranded recognition arms because helices I and III remain base paired after self-cleavage. Engineering trans-acting hammerhead with this feature (base paired recognition arms), did not block trans activity in vitro, and actually enhanced the specificity of the ribozyme.[59] Another recent improvement in hammerhead formulations includes the addition of a poly A tail at the 3' end of the ribozyme. The poly A may help to destroy hammerhead targets by recruiting RNA helicases that help to unfold strong secondary structures in the target RNA.[57]

Loop-Loop and Loop-Helix Interactions

Studies in vitro allowed defining the catalytic core of the hammerhead and the nucleotides required to catalyze a cleavage reaction. However, those studies ignored sequence elements outside the catalytic core, which are required for efficient activity in vivo where magnesium concentrations are much lower. Since in vitro studies mostly use the trans format any positive effect of the loops I and III on the cleavage reaction are lost in this formulation. In the three dimensional structure of the hammerhead, loop I and loop II are very close and nonWatson Crick base pairs could mediate loop I-loop II interactions[60] (Fig. 3A). Accordingly, self-cleaving motifs containing the loops I and II found in nature efficiently self-cleaved in vivo, while self-cleaving motifs lacking the natural loops did not.[60,61] This study suggests that the difficulties in applying the trans reaction in vivo to cleave cellular RNAs could be solved if the interaction between loop 1 and loop II could be modeled in trans.

Transcripts of the newt satellite 2 DNA also possess a hammerhead self-cleaving motif (Fig. 3B) characterized by a very short helix III (only two nucleotides long) and an internal loop in helix I.[62,63] This internal loop is required for efficient cleavage by the newt hammerhead possibly by mediating interactions with loop II (Fig. 3B). Intriguingly, ribozymes cloned from different species of salamanders differ in the identity of the nucleotides forming the internal loop. An analyses of chimeric transcripts of hammerhead from different salamanders indicated that each internal loop is functional when combined with the helix II of the same species. Mutational analysis confirmed that there are specific sequence and/or structural requirements in helix II that mediates interaction with the internal loop in helix I.[64] By analogy these helix II-internal loop interactions in the newt hammerheads may substitute the loop-loop interactions found in the hammerheads from plant small circular RNAs (Fig. 3B).

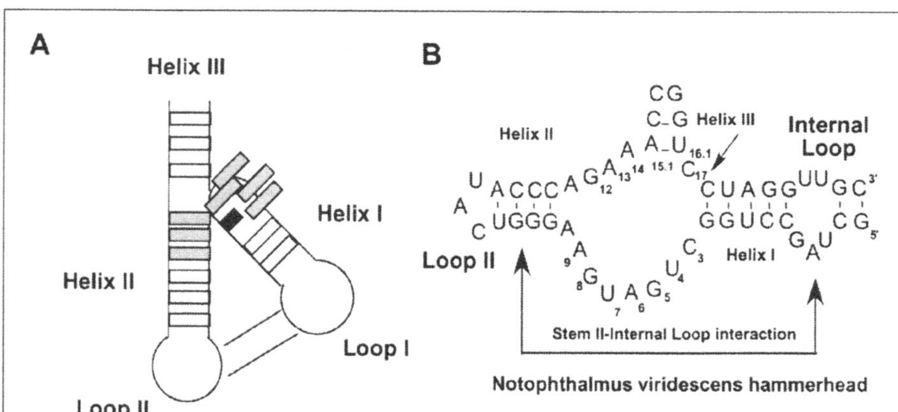

Figure 3. Peripheral elements of the hammerhead motif. A) Loop-loop interactions present in all natural self-cleaving hammerheads found in viroids, virusoids and satellite RNAs. Two interloop contacts one non Watson Crick and one trans-Watson Crick have been proposed.[60] B) Extended hammerhead motif from the satellite 2 DNA transcripts of the newt *Notophthalmus viridescens*.[61] Note that the loop I-loop II interactions found in viroids might be substituted by stem II-internal loop interactions in the newt.

Discovery of Self-Cleaving Transcripts from the Smα SINEs in Schistosoma

The discovery of hammerhead ribozymes first in small plant circular RNAs and later in the satellite DNA of several species of salamanders suggested a multiple and independent origin for this simple RNA motif. In plant circular RNAs the hammerhead is used to process the products of rolling circle replication, while in salamanders their function is not yet well-known. Since salamanders contains both multimeric and monomer transcripts of the satellite 2 DNA, it was proposed that the hammerhead catalyzed the processing of multimeric copies transcribed from tandemly arrayed satellite 2 DNA.[62]

The hammerhead ribozyme motif is simple enough to be described as a combination of helices and single stranded nucleotides. Ferbeyre and colleagues used the search engine RNAMOT[65] to look for sequences containing hammerhead domains in the GenBank.[41,66] This search, not surprisingly, found the known hammerhead domains in plant small circular RNAs and several newt species. It also found novel hammerhead RNA motifs in different families of repetitive DNA including the Smα family from *Schistosoma mansoni* (Fig. 4A) and the pDo500 family from *Dolichopoda* cave crickets.[67] Related sequences to Smα also possessing hammerhead ribozymes motifs were subsequently cloned from *S. haematobium* (Shα), *S. douthitti* (Sdα)[41] and *S. japonicum* (Sjα).[42]

Although the hammerhead motif found in plants, newts, schistosomes and crickets all possess the essential nucleotides required for self-cleavage, they are all embedded in different sequence contexts and display considerable variation in the nonessential nucleotide positions (Fig. 4B). This suggests that the small hammerhead ribozyme motif independently arose in different sequences. Accordingly, the hammerhead motif was easily recovered by Salehi-Ashtiani and Szostak from a randomized pool of RNA molecules after in vitro selection for self-cleavage.[68]

The hammerhead motif of the Smα repeats, like those found in the newt, does not possess the loop I. Instead, helix I bears an internal loop that is highly conserved among all hammerhead motifs found in different species of schistosomes. One important difference between the newt and the *Schistosoma* hammerhead is found in helix III. While the newt hammerhead displays a very short, two nucleotides long, helix III, in Schistosoma, this helix is extended to

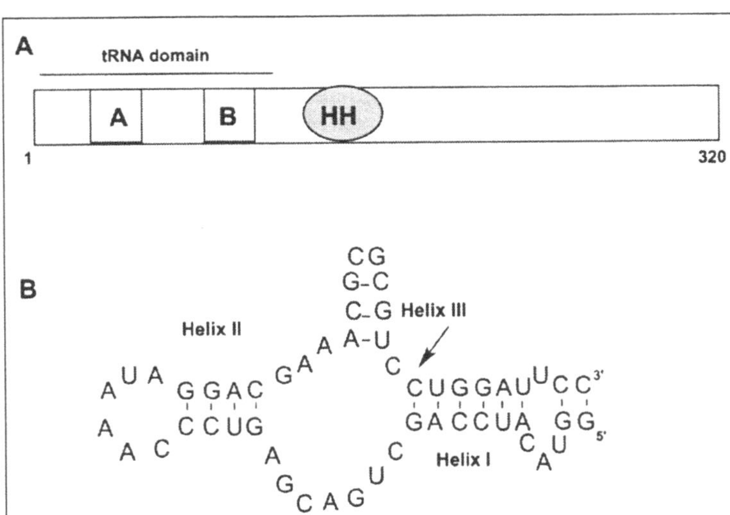

Figure 4. The Smα repeats from *Schistosoma mansoni* encode hammerhead ribozymes. A) Domain organization of the Smα repeats. The 5' tRNA domain contains boxes A and B for transcription by RNA polymerase III. The hammerhead self-cleaving motif is in the middle region.

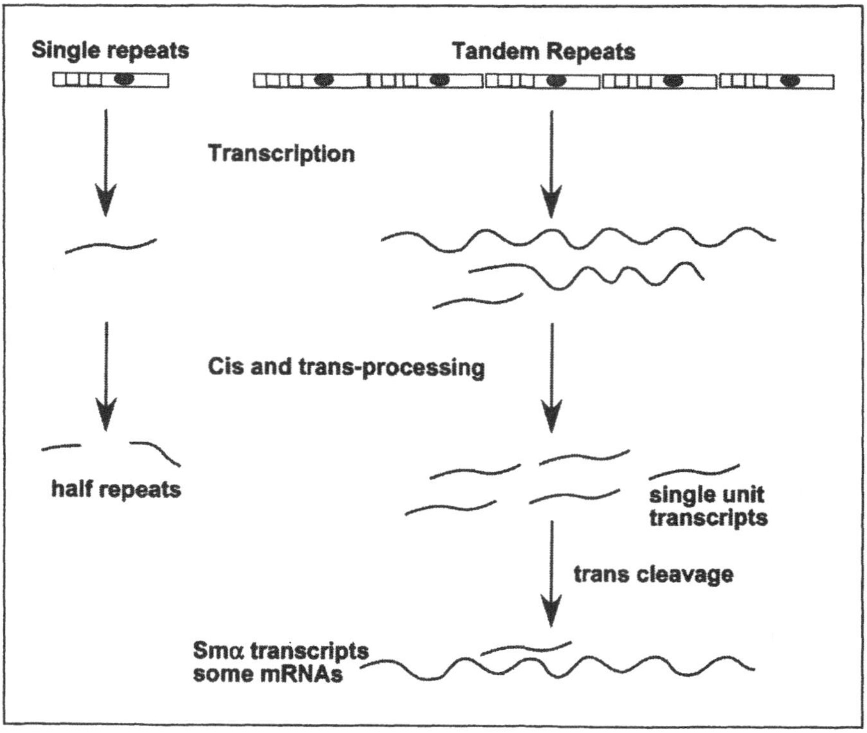

Figure 5. Transcription and processing of transcripts from Smα repeats. Smα repeats are transcribed from tandem repeats or solitary repeats in the *S. mansoni* chromosomes. A transcript from solitary repeats self-cleave via hammerhead ribozymes into half repeats. A Multimeric transcript from tandem repeats self-cleave also via hammerhead domains into single unit repeats. However, since many repeats contain mutated hammerhead ribozymes cleavage must occur in trans, most probably from the active single repeats that accumulate in Schistosoma tissues.

three nucleotides. As a consequence the hammerhead motif from Schistosoma self-cleaves with very high efficiency. At 37°C, the rate of cleavage could not be measured because most transcripts bearing the ribozyme essentially self-cleaved in less than a minute. To measure the kinetic parameters of the reaction, the temperature had to be reduced to 30°C and in those conditions the Schistosoma hammerhead was still more active than the newt hammerhead. The Schistosoma hammerhead is also remarkably stable. High rate of cleavage was observed at 70°C. Accordingly, when introduced into the extreme thermophile *Thermos thermophilus*, Smα repeats self cleaved with high efficiency.[69]

Smα repeats are highly expressed in *S. mansoni*. The major species in Northern blots corresponds to unit length repeats of around 320 nucleotides. Multiples of these repeats of higher molecular weight were also detected. In schistosomules (the larval form that develops from cercariae in infected hosts), Northern analysis revealed the presence of Smα fragments of around 160 bp, which is the length of the fragments predicted to arise from the cleavage of a single 320 bp repeat. Although different models could account for this expression pattern, it was proposed that transcription from tandemly arrayed repeats lead to multimeric transcripts that are then processed by the hammerhead motif into unit length monomers.[41] The pattern in schistosomules is consistent with the transcription of isolated repeats that become half repeats after hammerhead-mediated self-cleavage (Fig. 5) or with the presence of DNA versions of the half repeats that contain promoter elements.

Transcription and Propagation of Smα Repeats

Transcription of Smα repeats most probably occurs from the polymerase III Box A and Box B elements present in the tRNA domain of the repeats (Fig. 4A). In salamanders, hammerhead containing repeats are transcribed from polymerase II promoters similar to those in small nuclear RNAs.[70] Once more, the Smα repeats of schistosomes and the sat 2 repeats in the newt seem to have no evolutionary relationship. Since polymerase III promoters are constitutive, Smα repeats must be transcribed in every cell and in every developmental stage of schistosomes.

At the present, we can only speculate about how Smα repeats have invaded schistosomes' genomes. In view of the fact that Smα repeats do not encode any protein they most probably used the reverse transcriptase from an endogenous retrotransposon for propagation. Hence, Smα and related repeats are nonautonomous retrotransposons or SINEs, like human Alu sequences. Some Smα repeats contain three adenines immediately after the hammerhead cleavage site. This may be the remains of polyadenylation after cleavage and an indirect proof of their origin via reverse transcription and integration in the genome. In fact, sequence analysis of *S. haematobium* Shα repeats revealed that most isolates contained the triple A insertion after the cleavage site.[41]

The preponderance of single unit transcripts of Smα repeats was an unexpected surprise. Sequence analysis of multiple hammerhead-containing Smα repeats revealed that most hammerhead motifs possessed disabling mutations in the catalytic core. This means that in order to achieve complete processing of multimeric transcripts one of the following possibilities must occur.

1. Trans cleavage of inactive repeats.
2. Distal-cis cleavage where an active hammerhead from an unprocessed repeat cleave an inactive one before self-cleaving itself.
3. Only repeats with active domain are transcribed.

Distal cis-cleavage has been proposed before to explain the cleavage of multimeric RNAs possessing hammerhead motifs with very short helix III. In this case each hammerhead is compromised in its own self-cleavage due to the short and unstable helix III as seen in salamanders.[62] Forming the so called double hammerhead structures between adjacent motifs ensure the formation of stable structures that catalyze the cleavage of both motifs.[45] Distal cis-cleavage seemed less likely for Smα transcripts because their self-cleaving motifs are very active. That is, they will most probably self-cleave before folding into a conformation that permits the catalytic part of a repeat to base pair with the nucleotides flanking the cleavage site of another repeat possessing disabling mutations. It is also very difficult to imagine a mechanism that restricts transcription to active repeats. Polymerase III elements are present in all repeats and contain no mutations. Therefore Smα repeats may possess the ability to catalyze RNA cleavage in trans, a property that made them very attractive to develop RNA therapeutics.

Trans Reactions Catalyzed by the Smα Repeats

The hammerhead motif can be reengineered in two different ways to cleave RNA targets in trans. In the I-III format, most of the essential nucleotides are kept in the catalytic part and the substrate only provides the cleavage site, where only U16.1 and H17 are required (H is C, A or U). This is the format commonly used by most studies that employ the hammerhead ribozyme to cleave other cellular RNAs.[47] In the I-II format the substrate provides, in addition of the UH nucleotides at the cleavage site, part of the helix II and the essential nucleotides G12A13A14A15.1. Such format has been previously shown to be active in vitro.[71] In Smα repeats, nucleotides G12-A15 are rarely inactivated. This allows to single-unit transcripts to use the I-II format to cleave repeats with disabling mutations in the essential sequence from C3-A9.

Even more interesting is the fact that some *S. mansoni* protein coding genes contain fragments of the Smα sequences that are not sufficient for self-cleavage but could be cleaved in trans using the I-II format. The presence of Smα sequences in other genes is not a biological oddity of Schistosomes. In the human genome there are 2400 proteins containing Alu or L1 sequences.[72] The gene coding for synaptobrevin in *S. mansoni* contains a fragment of sequence almost identical to a fragment of Smα repeats (Fig. 6A). It was shown in vitro that single-unit transcripts from

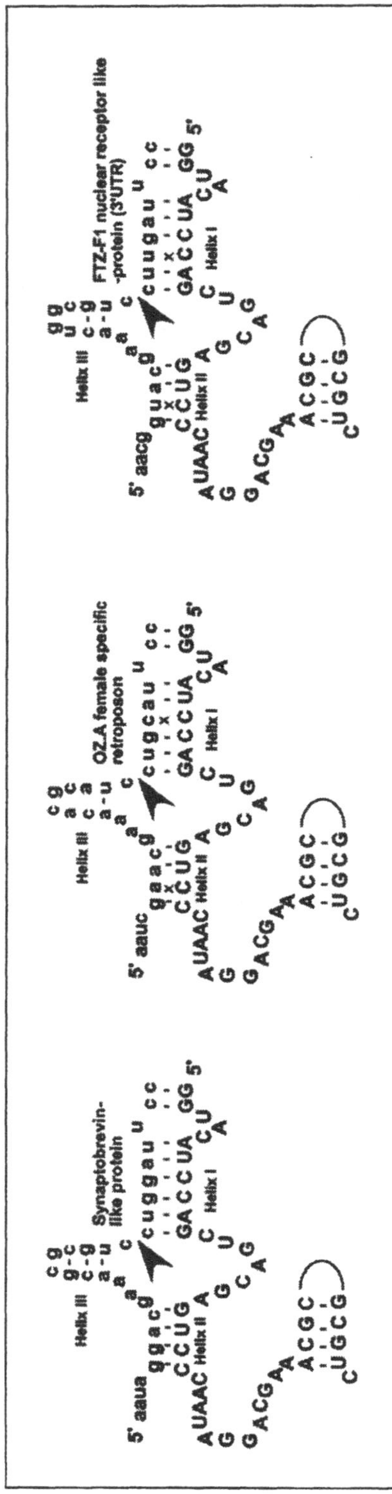

Figure 6. Trans Reactions Catalyzed by Smα Ribozyme. A) Synaptobrevin-like protein (SMU30291, intron, from 1245 to 1273. B) OZ.A female specific retrotransposon (SMU12442). C) FTZ-F1 nuclear receptor-like protein, 3'UTR, from 23645 to 23668.

Smα repeats cleaved *Schistosoma* synaptobrevin mRNA.[41] In addition transcripts form the retro-transposon OZ.A were predicted to be cleaved by the Smα ribozyme (Fig. 6B).[41] Another candidate target for cleavage in trans by the Smα ribozyme is the 3' UTR of the FTZ-F1 nuclear receptor-like protein. The recognition of the OZ.A retrotransposon and the FTZ-F1 3'UTR sequences by any of the known Smα repeats would include two mismatches one in helix I and one in helix II (Fig. 6B). Cleavage may still occur despite these mismatches but it is possible that sequence variation among different Smα repeats may help to recognize different substrates.

Concluding Remarks

The self-cleaving ability of the Smα family of repeated sequences in *S. mansoni* and of similar families found in *S. haematobium, S. japonicum and S. douthitti* raises interesting questions about their possible functions. Repeated sequences from several newt species and the cave cricket also possess self-cleaving domains of the hammerhead type. Since there is no other homology between these different families of repeated sequences, the most plausible scenario is that of independent origin of the hammerhead motif.

Two different functions have been proposed for self-cleaving repeats.[41] As mentioned before they can be used to cleave other mRNAs in trans to control gene expression. The examples of synaptobrevin-like protein and the FTZ-F1 mRNAs are consistent with this proposal but further studies are required to prove whether these trans reactions occur in vivo and influence the expression of the target genes. In the newt, the hammerhead containing transcripts were found as part of 12S ribonucleoprotein complexes[73] suggesting a functional role for the ribozymes, perhaps in catalyzing the cleavage *in* trans of other cellular RNAs.

The second function proposed for the self-cleaving reaction is that of self-control of the transposition abilities of repeated sequences. Self-cleavage might initiate the destruction of the transcripts coming from repeated sequences; therefore limiting the damage that retrotransposition can impose to the host. The retrotransposon *micropia* from *Drosophila* uses a related strategy to limit its expansion.[74] *Micropia* encodes a testis specific antisense RNA complementary to the sequence of its own reverse transcriptase. The Smα and related ribozymes could accomplish a similar function by restricting the propagation of SINES. Therefore hammerhead mediated RNA cleavage is another mechanism to control retrotransposon propagation different from the known processes of silencing, targeted integration and premature polyadenylation. Conceptually self-cleavage is similar to premature polyadenylation but with the advantage of generating active ribozymes that can act in trans on other transcripts. Perhaps this self cleaving strategy is used more frequently than currently recognized by mobile elements. Therefore, other self-cleavage motifs wait to be discovered among the many families of repeated sequences that have invaded eukaryotic genomes.

References

1. Kazazian Jr HH. Mobile elements: Drivers of genome evolution. Science 2004; 303(5664):1626-1632.
2. Lazcano A, Miller SL. The origin and early evolution of life: Prebiotic chemistry, the preRNA world, and time. Cell 1996; 85(6):793-798.
3. Xiong Y, Eickbush TH. Origin and evolution of retroelements based upon their reverse transcriptase sequences. EMBO J 1990; 9(10):3353-3362.
4. Garfinkel DJ, Boeke JD, Fink GR. Ty element transposition: Reverse transcriptase and virus-like particles. Cell 1985; 42(2):507-517.
5. Paces J, Pavlicek A, Zika R et al. HERVd: The human endogenous retroviruses database: Update. Nucleic Acids Res 2004; 32(Database issue):D50.
6. Kim A, Terzian C, Santamaria P et al. Retroviruses in invertebrates: The gypsy retrotransposon is apparently an infectious retrovirus of Drosophila melanogaster. Proc Natl Acad Sci USA 1994; 91(4):1285-1289.
7. Laha T, Loukas A, Verity CK et al. Gulliver, a long terminal repeat retrotransposon from the genome of the oriental blood fluke Schistosoma japonicum. Gene 2001; 264(1):59-68.
8. Luan DD, Korman MH, Jakubczak JL et al. Reverse transcription of R2Bm RNA is primed by a nick at the chromosomal target site: A mechanism for non-LTR retrotransposition. Cell 1993; 72(4):595-605.

9. Kazazian Jr HH, Goodier JL. LINE drive. Retrotransposition and genome instability. Cell 2002; 110(3):277-280.
10. Jakubczak JL, Xiong Y, Eickbush TH. Type I (R1) and type II (R2) ribosomal DNA insertions of Drosophila melanogaster are retrotransposable elements closely related to those of Bombyx mori. J Mol Biol 1990; 212(1):37-52.
11. Malik HS, Eickbush TH. The RTE class of non-LTR retrotransposons is widely distributed in animals and is the origin of many SINEs. Mol Biol Evol 1998; 15(9):1123-1134.
12. Mol CD, Kuo CF, Thayer MM et al. Structure and function of the multifunctional DNA-repair enzyme exonuclease III. Nature 1995; 374(6520):381-386.
13. Eickbush TH. Telomerase and retrotransposons: Which came first? Science 1997; 277(5328):911-912.
14. Deininger PL, Batzer MA. Mammalian retroelements. Genome Res 2002; 12(10):1455-1465.
15. Daniels GR, Deininger PL. Repeat sequence families derived from mammalian tRNA genes. Nature 1985; 317(6040):819-822.
16. Ullu E, Tschudi C. Alu sequences are processed 7SL RNA genes. Nature 1984; 312(5990):171-172.
17. Yoder JA, Walsh CP, Bestor TH. Cytosine methylation and the ecology of intragenomic parasites. Trends Genet 1997; 13(8):335-340.
18. Walsh CP, Bestor TH. Cytosine methylation and mammalian development. Genes Dev 1999; 13(1):26-34.
19. Schramke V, Allshire R. Hairpin RNAs and retrotransposon LTRs effect RNAi and chromatin-based gene silencing. Science 2003; 301(5636):1069-1074.
20. Lin R, Avery L. RNA interference. Policing rogue genes. Nature 1999; 402(6758):128-129.
21. Sijen T, Plasterk RH. Transposon silencing in the Caenorhabditis elegans germ line by natural RNAi. Nature 2003; 426(6964):310-314.
22. Sandmeyer SB, Hansen LJ, Chalker DL. Integration specificity of retrotransposons and retroviruses. Annu Rev Genet 1990; 24:491-518.
23. SanMiguel P, Tikhonov A, Jin YK et al. Nested retrotransposons in the intergenic regions of the maize genome. Science 1996; 274(5288):765-768.
24. Han JS, Szak ST, Boeke JD. Transcriptional disruption by the L1 retrotransposon and implications for mammalian transcriptomes. Nature 2004; 429(6989):268-274.
25. Perepelitsa-Belancio V, Deininger P. RNA truncation by premature polyadenylation attenuates human mobile element activity. Nat Genet 2003; 35(4):363-366.
26. Nigumann P, Redik K, Matlik K et al. Many human genes are transcribed from the antisense promoter of L1 retrotransposon. Genomics 2002; 79(5):628-634.
27. Ferrigno O, Virolle T, Djabari Z et al. Transposable B2 SINE elements can provide mobile RNA polymerase II promoters. Nat Genet 2001; 28(1):77-81.
28. Bailey JA, Carrel L, Chakravarti A et al. Molecular evidence for a relationship between LINE-1 elements and X chromosome inactivation: The Lyon repeat hypothesis. Proc Natl Acad Sci USA 2000; 97(12):6634-6639.
29. Lyon MF. X-chromosome inactivation: A repeat hypothesis. Cytogenet Cell Genet 1998; 80(1-4):133-137.
30. Erlandsson R, Wilson JF, Paabo S. Sex chromosomal transposable element accumulation and male-driven substitutional evolution in humans. Mol Biol Evol 2000; 17(5):804-812.
31. Chu WM, Ballard R, Carpick BW et al. Potential Alu function: Regulation of the activity of double-stranded RNA-activated kinase PKR. Mol Cell Biol 1998; 18(1):58-68.
32. Rubin CM, Kimura RH, Schmid CW. Selective stimulation of translational expression by Alu RNA. Nucleic Acids Res 2002; 30(14):3253-3261.
33. Teng SC, Kim B, Gabriel A. Retrotransposon reverse-transcriptase-mediated repair of chromosomal breaks. Nature 1996; 383(6601):641-644.
34. Morrish TA, Gilbert N, Myers JS et al. DNA repair mediated by endonuclease-independent LINE-1 retrotransposition. Nat Genet 2002; 31(2):159-165.
35. Simpson AJ, Sher A, McCutchan TF. The genome of Schistosoma mansoni: Isolation of DNA, its size, bases and repetitive sequences. Mol Biochem Parasitol 1982; 6(2):125-137.
36. Copeland CS, Brindley PJ, Heyers O et al. Boudicca, a retrovirus-like long terminal repeat retrotransposon from the genome of the human blood fluke Schistosoma mansoni. J Virol 2003; 77(11):6153-6166.
37. Drew AC, Brindley PJ. A retrotransposon of the nonlong terminal repeat class from the human blood fluke Schistosoma mansoni. Similarities to the chicken-repeat-1-like elements of vertebrates. Mol Biol Evol 1997; 14(6):602-610.
38. Drew AC, Minchella DJ, King LT et al. SR2 elements, nonlong terminal repeat retrotransposons of the RTE-1 lineage from the human blood fluke Schistosoma mansoni. Mol Biol Evol 1999; 16(9):1256-1269.

39. Laha T, Brindley PJ, Smout MJ et al. Reverse transcriptase activity and untranslated region sharing of a new RTE-like, nonlong terminal repeat retrotransposon from the human blood fluke, Schistosoma japonicum. Int J Parasitol 2002; 32(9):1163-1174.
40. Spotila LD, Hirai H, Rekosh DM et al. A retroposon-like short repetitive DNA element in the genome of the human blood fluke, Schistosoma mansoni. Chromosoma 1989; 97(6):421-428.
41. Ferbeyre G, Smith JM, Cedergren R. Schistosome satellite DNA encodes active hammerhead ribozymes. Mol Cell Biol 1998; 18(7):3880-3888.
42. Laha T, McManus DP, Loukas A et al. Sjalpha elements, short interspersed element-like retroposons bearing a hammerhead ribozyme motif from the genome of the oriental blood fluke Schistosoma japonicum. Biochim Biophys Acta 2000; 1492(2-3):477-482.
43. Cech TR, Bass BL. Biological catalysis by RNA. Annu Rev Biochem 1986; 55:599-629.
44. Altman S. Ribonuclease P. Postscript. J Biol Chem 1990; 265(33):20053-20056.
45. Symons RH. Small catalytic RNAs. Annu Rev Biochem 1992; 61:641-671.
46. Doudna JA, Cech TR. The chemical repertoire of natural ribozymes. Nature 2002; 418(6894):222-228.
47. Bratty J, Chartrand P, Ferbeyre G et al. The hammerhead RNA domain, a model ribozyme. Biochim Biophys Acta 1993; 1216(3):345-359.
48. Ruffner DE, Stormo GD, Uhlenbeck OC. Sequence requirements of the hammerhead RNA self-cleavage reaction. Biochemistry 1990; 29(47):10695-10702.
49. Ambros S, Flores R. In vitro and in vivo self-cleavage of a viroid RNA with a mutation in the hammerhead catalytic pocket. Nucleic Acids Res 1998; 26(8):1877-1883.
50. Perriman R, Delves A, Gerlach WL. Extended target-site specificity for a hammerhead ribozyme. Gene 1992; 113(2):157-163.
51. McCall MJ, Hendry P, Jennings PA. Minimal sequence requirements for ribozyme activity. Proc Natl Acad Sci USA 1992; 89(13):5710-5714.
52. Scott WG, Finch JT, Klug A. The crystal structure of an all-RNA hammerhead ribozyme: a proposed mechanism for RNA catalytic cleavage. Cell 1995; 81(7):991-1002.
53. Hendry P, McCall M. Unexpected anisotropy in substrate cleavage rates by asymmetric hammerhead ribozymes. Nucleic Acids Res 1996; 24(14):2679-2684.
54. Hammann C, Lilley DM. Folding and activity of the hammerhead ribozyme. Chembiochem 2002; 3(8):690-700.
55. De la Pena M, Flores R. An extra nucleotide in the consensus catalytic core of a viroid hammerhead ribozyme: Implications for the design of more efficient ribozymes. J Biol Chem 2001; 276(37):34586-34593.
56. Stage-Zimmermann TK, Uhlenbeck OC. A covalent crosslink converts the hammerhead ribozyme from a ribonuclease to an RNA ligase. Nat Struct Biol 2001; 8(10):863-867.
57. Kawasaki H, Taira K. Identification of genes by hybrid ribozymes that couple cleavage activity with the unwinding activity of an endogenous RNA helicase. EMBO Rep 2002; 3(5):443-450.
58. Paddison PJ, Hannon GJ. RNA interference: The new somatic cell genetics? Cancer Cell 2002; 2(1):17-23.
59. Ohmichi T, Kool ET. The virtues of self-binding: High sequence specificity for RNA cleavage by self-processed hammerhead ribozymes. Nucleic Acids Res 2000; 28(3):776-783.
60. Khvorova A, Lescoute A, Westhof E et al. Sequence elements outside the hammerhead ribozyme catalytic core enable intracellular activity. Nat Struct Biol 2003; 10(9):708-712.
61. De la Pena M, Gago S, Flores R. Peripheral regions of natural hammerhead ribozymes greatly increase their self-cleavage activity. EMBO J 2003; 22(20):5561-5570.
62. Epstein LM, Gall JG. Self-cleaving transcripts of satellite DNA from the newt. Cell 1987; 48(3):535-543.
63. Pabon-Pena LM, Zhang Y, Epstein LM. Newt satellite 2 transcripts self-cleave by using an extended hammerhead structure. Mol Cell Biol 1991; 11(12):6109-6115.
64. Garrett TA, Pabon-Pena LM, Gokaldas N et al. Novel requirements in peripheral structures of the extended satellite 2 hammerhead. RNA 1996; 2(7):699-706.
65. Laferriere A, Gautheret D, Cedergren R. An RNA pattern matching program with enhanced performance and portability. Comput Appl Biosci 1994; 10(2):211-212.
66. Ferbeyre G, Bourdeau V, Pageau M et al. Distribution of hammerhead and hammerhead-like RNA motifs through the GenBank. Genome Res 2000; 10(7):1011-1019.
67. Rojas AA, Vazquez-Tello A, Ferbeyre G et al. Hammerhead-mediated processing of satellite pDo500 family transcripts from Dolichopoda cave crickets. Nucleic Acids Res 2000; 28(20):4037-4043.
68. Salehi-Ashtiani K, Szostak JW. In vitro evolution suggests multiple origins for the hammerhead ribozyme. Nature 2001; 414(6859):82-84.

69. Vazquez-Tello A, Castan P, Moreno R et al. Efficient trans-cleavage by the Schistosoma mansoni SMalpha1 hammerhead ribozyme in the extreme thermophile Thermus thermophilus. Nucleic Acids Res 2002; 30(7):1606-1612.
70. Coats SR, Zhang Y, Epstein LM. Transcription of satellite 2 DNA from the newt is driven by a snRNA type of promoter. Nucleic Acids Res 1994; 22(22):4697-4704.
71. Chartrand P, Harvey SC, Ferbeyre G et al. An oligodeoxyribonucleotide that supports catalytic activity in the hammerhead ribozyme domain. Nucleic Acids Res 1995; 23(20):4092-4096.
72. Li WH, Gu Z, Wang H et al. Evolutionary analyses of the human genome. Nature 2001; 409(6822):847-849.
73. Luzi E, Eckstein F, Barsacchi G. The newt ribozyme is part of a riboprotein complex. Proc Natl Acad Sci USA 1997; 94(18):9711-9716.
74. Lankenau S, Corces VG, Lankenau DH. The Drosophila micropia retrotransposon encodes a testis-specific antisense RNA complementary to reverse transcriptase. Mol Cell Biol 1994; 14(3):1764-1775.
75. Branch AD, Robertson HD. A replication cycle for viroids and other small infectious RNA's. Science 1984; 223(4635):450-455.

CHAPTER 2

Mobile Genetic Elements of Malaria Vectors and Other Mosquitoes

Zhijian Tu* and Song Li

Abstract

Mosquitoes are important vectors of a number of disease agents including malarial and filarial parasites as well as many types of viruses. A wide spectrum of both RNA-mediated and DNA-mediated transposable elements (TEs) have been discovered in the African malaria mosquito, *Anopheles gambiae*, and other mosquito species. Mosquito TEs are potentially useful as tools for genetic manipulation, or as markers for population studies. Such genetic and population studies are relevant to the long-term strategy to control mosquito-borne infectious diseases through the introduction and spread of refractory mosquitoes in natural populations. Here we present an overview of the characteristics of mosquito TEs. We highlight several mosquito TEs that appear to be transpositionally active and a few newly discovered elements that offer interesting perspectives on TE diversity, classification, and transposition. TE display, a PCR-based genome scan method, is also described in the context of uncovering polymorphism at TE insertion sites and detecting new transposition events. We conclude the chapter outlining questions that are of fundamental and applied significance in mosquito TE research and providing promising new research avenues in light of a growing number of mosquito genome projects and systems biology methods.

Introduction

Transposable elements (TEs) are mobile genetic elements that are widely distributed in bacteria, archaea, and eukaryotes.[1] TEs are generally thought to replicate and spread in their host genomes as primarily "selfish" genetic units.[2,3] However, recent genomic and experimental data suggest that the "selfish" TEs could evolve a spectrum of relationships with their hosts, ranging from "junk parasites" to "molecular symbionts".[4,5] Some TEs have been shown to perform important biological functions in their hosts. For example, the RAG1 and RAG2 proteins, which are enzymes that are involved in V(D)J recombination in jawed vertebrates, originate from an ancient transposase.[6] In fact, a recent analysis has identified a *Transib* transposon as the modern-day relative of RAG1.[7] V(D)J recombination is very important to the vertebrate immune system as it significantly contributes to the diversity of antibodies and T-cell receptors.

There are two classes of TEs that utilize different transposition mechanisms. Class I RNA-mediated TEs synthesize cDNA molecules that are integrated in the genome. The cDNA molecules are reverse transcribed from TE transcripts, which serve as the retrotransposition intermediates.[8] Class II DNA-mediated TEs transpose directly from DNA to DNA and they do not need an RNA intermediate.[1] Instead, DNA-mediated TEs transpose through either "cut and paste" or "rolling circle" mechanisms. In most cases, insertion of both classes of TEs

*Corresponding Author: Zhijian Tu—Department of Biochemistry, Virginia Tech, Blacksburg, VA 24061, USA. Email: jaketu@vt.edu

Mobile Genetic Elements in Metazoan Parasites, edited by Paul J. Brindley.

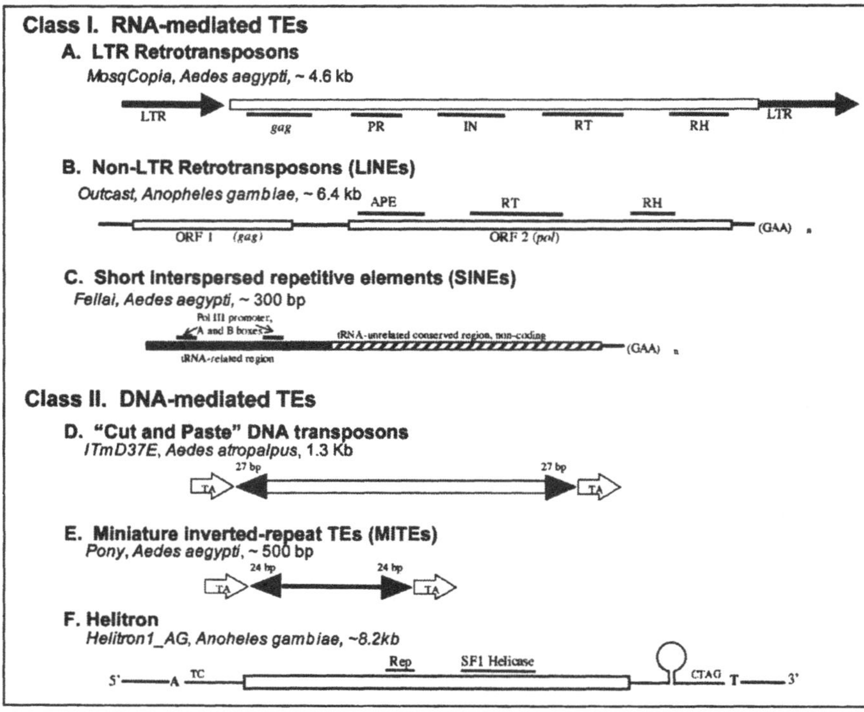

Figure 1. Structural characteristics of TE representatives in mosquitoes. Representatives of RNA-mediated TEs are shown from three major groups, long terminal repeat (LTR) retrotransposons (A), non-LTR retrotransposons (B), and short interspersed elements (SINEs, C). Representatives of DNA-mediated TEs are shown from three major groups, cut and paste DNA transposons (D), miniature inverted-repeat TEs (MITEs, E), and *Helitrons* (F). The name of each representative element, its host species, and its approximate length are shown in the heading. The elements are not drawn to scale. Open reading frames (ORFs) are shown as open boxes. For DNA-mediated TEs, open arrows indicate target site duplications (TSDs). Filled triangles indicate terminal inverted repeats. The lengths of these inverted repeats are marked. Note that *Helitrons* have neither TSDs nor TIRs. The terminal sequences and the one base flanking sequences of *Helitrons* are shown. *env*: envelope protein; *gag*: group-associated antigen, or group-specific antigen; IN: Integrase; LINEs: Long interspersed elements; LTR: Long terminal repeat; PR: Protease; RH: RNase H; RT: Reverse transcriptase. References for information on these TEs are as following. *MosqCopia* (Z. Tu and M. Yoshiyama, unpublished data; GenBank AY009101); *Outcast,*[37] *Feilai,*[56] *ITmD37E,*[35] *Pony,*[63] *Helitron.*[15]

causes the duplication of the insertion target. Such target site duplications (TSDs) flank the actual TE sequence and serve as a hallmark of TE insertion.

RNA-mediated TEs include long terminal repeat (LTR) retrotransposons, non-LTR retrotransposons, and short interspersed elements (SINEs). Non-LTR retrotransposons are also referred to as retroposons or long interspersed elements (LINEs). The structural features of the three groups of RNA-mediated TEs are illustrated in Figure 1A-C. The LTRs in the LTR retrotransposons are normally 200 to 500 bp long and provide promoter sequences and transcription termination signals.[8] LTR retrotransposons encode a polymerase (*pol*) protein that contains reverse transcriptase (RT), ribonuclease H (RNase H), protease (PR), and integrase (IN) domains that are important for their retrotransposition. The RT domain catalyzes reverse transcription and its sequence is used for phylogenetic classification of LTR retrotransposons.[8] Like the LTR retrotransposons, most non-LTR retrotransposons also have a *pol*-like protein that includes a RT domain that is essential for retrotransposition. Some non-LTR elements also have an RNase H and/or endonuclease domain encoded in the *pol* open reading frame. In addition to the *pol*-like protein, many non-LTR elements encode a protein related to the retroviral group-associated antigen (*gag*) protein. Non-LTR retrotransposons are transcribed from internal Pol II promoters. Some contain AATAAA polyadenylation signals, poly (A) tails, or simple tandem repeats at their 3' ends. Target Primed Reverse Transcription has been proposed as the mechanism of retrotransposition for the non-LTR elements.[9,10] SINEs, the third group of Class I TEs, are generally between 100 to 500 bp long and they do not have any coding potential. It has been shown that some SINEs retrotranspose by "borrowing" the machinery from autonomous non-LTR retrotransposons, and that this process may be facilitated by the presence of similar sequences or structures at the 3' ends of a SINE and its "partner" non-LTR retrotransposon.[11-14] Unlike non-LTR retrotransposons that use internal Pol II promoters, SINE transcription is directed from Pol III promoters that are similar to those found in small RNA genes.

As shown in Figure 1D-F, DNA-mediated TEs include cut and paste DNA transposons, miniature inverted-repeat TEs (MITEs), and *Helitrons*.[15,16] DNA transposons such as *P*, *Hermes*, *piggyBac* and *MosI* are typically characterized by 10-200 bp terminal inverted-repeats (TIRs), flanking one or more open reading frames that encode a transposase. They usually transpose by a cut-and-paste mechanism and their copy number may be increased through a repair mechanism.[1,17] MITEs are short nonautonomous TEs that share common structural characteristics such as TIRs, small size, and lack of coding sequence.[18] MITEs are thought to transpose by "borrowing" the machinery of autonomous DNA transposons, taking advantage of shared TIRs.[19-21] An alternative hypothesis suggests that they may transpose by a DNA hairpin intermediate produced from the folding-back of single-stranded DNA during replication.[22] *Helitrons*, the third group of Class II TEs, appear to use a rolling-circle mechanism of transposition.[15,16,23] Instead of a cut and paste transposase, *Helitrons* encode proteins similar to helicases, ssDNA-binding proteins, and replication initiation proteins. These proteins facilitate the rolling-circle replication of *Helitrons*, a mechanism previously described for the bacterial *IS91* transposons.[24]

Because of the potential utility of TEs as genetic tools to manipulate mosquitoes and other important disease vectors, molecular analysis of mosquito TEs has attracted significant attention. The report of the *An. gambiae* genome sequence and the near completion of the genome assembly of the yellow fever mosquito, *Aedes aegypti*, made it possible to systematically study mosquito TEs and offered a great opportunity for comparative analysis. There are a few recent reviews on insect TEs and their use as genetic tools.[25-27] In this chapter, we will review the different types of TEs in the African malaria mosquito, *An. gambiae* and other mosquitoes. We will focus on recent advances in mosquito TE research and provide a perspective for future investigations.

Overview of Mosquito TEs

A number of TEs have been discovered from *An. gambiae, Ae. aegypti* and other mosquitoes[28-35] before the availability of genome sequence data. The publication of the *An. gambiae* genome sequence assembly[36] included an initial survey of TEs. Several detailed characterizations of different groups of TEs in the *An. gambiae* genome have since been reported.[37-43] TEs occupy at least 16% of the euchromatic region in *An. gambiae*[36] and the non-LTR and LTR retrotransposon groups contain tremendously diverse TEs.[37,41] Initial analyses of the early version of the *Ae. aegypti* assembly (The Institute for Genomic Research, February, 2005) indicate that more than 40% of the genome consists of TE derived sequences (Z. Tu, unpublished data). This number is likely to increase as more careful investigations are completed and a better assembly becomes available. Non-LTRs, SINEs, and MITEs are the predominant types of TEs in *Ae. aegypti* with regard to copy number. *JuanA*, a non-LTR retrotransposon in *Ae. aegypti*, consists of more than 10,000 copies and occupies at least 3% of the entire genome. The majority of TEs reported in mosquitoes are from either *An. gambiae* or *Ae aegypti*. However, there are several reports of both classes of TEs in other mosquitoes and there are PCR surveys of TEs in a number of species.[33,44-47] All major groups of TEs in both classes have been found in mosquitoes. Figure 1 shows the characteristics of representatives of the major groups of RNA-mediated and DNA-mediated TEs. Tables 1 and 2

Table 1. RNA-mediated transposable elements in mosquitoes

Superfamily (Clade)	Element	Organism
I. LTR retrotransposons		
Ty1/copia	*Mosqcopia* (Z. Tu and M. Yoshiyama, unpublished), *Zebedee* [28]	*Aedes aegypti*
	copia-like [36], *mtanga* [48]	*Anopheles gambiae*
	AAcopia1_1, Copia_DM [38]	*Anopheles gambiae*
Ty3/gypsy	*Beagle, gypsy-like, Cruiser, Osvaldo, Springer* [36]	*Anopheles gambiae*
	Ozymandias [34]	*Anopheles gambiae*
	HMSbeagle_1, Blastopia_LTR,Roo_1, Invader2_1, Mdg3_DM, Stalker2_1, Quasimodo_1 [38]	*Anopheles gambiae*
	GYPSY18_AG - GYPSY72_AG [41]	*Anopheles gambiae*
BEL	*MosqNinja* (Z. Tu, unpublished)	*Aedes aegypti*
	Moose, Pao-like [36]	*Anopheles gambiae*
	Belly (M. Alvarez et al,. unpublished)	*Anopheles gambiae*
	Roo_1, Diver_1 [38]	*Anopheles gambiae*
	Indy[49]	*Anopheles gambiae*
II. Non-LTR retrotransposons		
R4	*Ag-R4_1* [37]	*Anopheles gambiae*
L1	*Ag-L1_1-5*[37]	*Anopheles gambiae*
L2	*Ag-L2_1-3* [37]	*Anopheles gambiae*
RTE	*JAM1* (GenBank Z86117)	*Aedes aegypti*
	Ag-Jammin_1-2 [37]	*Anopheles gambiae*
R1	*RT1, RT2, Ag-R1_1-11* [37,50]	*Anopheles gambiae*
	RTAg3, TRAg4, WaldoAg1,WaldoAg2, MinoAg1,R6Ag1,R6Ag2, R6Ag3, R7Ag1, R7Ag2, HidaAg1, NotoAg1, KagaAg1 [51]	*Anopheles gambiae*
	WRSRAs1, WRSRAa2 [51]	*Aedes aegypti*
LOA	*Lian* [52]	*Aedes aegypti*
I	*MosqI* [53]	*Aedes aegypti*
	Ag-I_1-7 [37]	*Anopheles gambiae*
Jockey	*JuanA* [30]	*Aedes aegypti*
	Ag-Jockey_1-13, Ag-Jen_1-12, [37]	*Anopheles gambiae*
CR1	*T1, Q, Ag-CR1_1-29* [29,37,54]	*Anopheles gambiae*
Loner	*Ag-Loner_1-3* [37]	*Anopheles gambiae*
Outcast	*Ag-Outcast_1-11* [37]	*Anopheles gambiae*
Not classified	*Sponge* [37]	*Anopheles gambiae*
	CM-gag [55]	*Culex pipiens*
III. SINEs		
tRNA-related SINEs	*Feilai* [56]	*Aedes aegypti*
	SINE200 [36]	*Anopheles gambiae*
	Twin-Sine [33]	*Culex pipiens*
	gecko [57]	*Aedes aegypti*
Unclassified SINEs?	*Maque* [58]	*Anopheles gambiae*

a. Classifications of LTR and non-LTR retrotransposons are according to Eickbush and Malik.[8]
b. The citation is not always the original report of a TE. In some cases a review or an article describing the phylogenetic placement of the TE is cited instead.
c. There are several PCR surveys of a few RNA-mediated TEs in mosquitoes that are not listed in the table.[45,46]
d. *Penelope*-like retrotransposons have also been found in *Aedes aegypti. Penelope* is a unique retrotransposon because it contains introns.
e. Preliminary genome analysis revealed a large number of LTR and non-LTR retrotransposons in *Ae. aegypti,* which are not listed here.

provide the names, classification, and characteristics of the two classes of TEs in mosquitoes. The two tables include substantial updates that were not discussed in previous reviews.[25,27]

Highlights and Recent Developments

Herves, an Active hAT-Like Transposon in An. gambiae

One exiting development in mosquito TE research is the report of an active DNA-mediated transposon in *An. gambiae.*[59] This element, named *Herves*, is a member of a subfamily within the *hobo-Ac-Tam3* (*hAT*) family. *Herves* was discovered during a whole-genome computational analysis for active DNA transposons. *Herves* was shown to be transpositionally active in mobility assays performed in *Drosophila melanogaster* S2 cells and developing embryos. *Herves* was also successfully used as a germline transformation vector in *D. melanogaster*. It remains to be determined whether *Herves* is currently active in *An. gambiae* populations. However, *Herves* insertion patterns in individuals of an East African population are at least consistent with it being active in mosquitoes. The few TEs that are used to transform mosquitoes are DNA transposons mainly from nonmosquito insects. *Herves* may be developed as an additional transformation vector for genetic manipulation of mosquitoes. Perhaps more importantly, if *Herves* is indeed actively transposing in *An. gambiae* populations, researchers will for the first time have the opportunity to study how active endogenous DNA transposons are regulated in mosquitoes and how this active element behaves in natural populations, which will be of great practical significance (see section on Practical Applications).

Gambol, a Novel Family of Transposons in the IS630-Tc1-Mariner *Superfamily*

Eukaryotic DNA transposon families *Tc1* and *mariner* and the bacterial *IS630* element and its relatives in prokaryotes and ciliates comprise a superfamily, the *IS630-Tc1-mariner* superfamily.[35,69-71] As shown in Table 2, the common motif of the transposase in this superfamily is a conserved D(Asp)DE(Glu) or DDD catalytic triad. *Tc1*-like elements identified in fungi, invertebrates and vertebrates all contain a DD34E motif. Analysis of the *An. gambiae* genome revealed a novel family of transposons named *gambol*, which uses a DD34E catalytic triad but are phylogenetically distinct from all *Tc1* transposons (Table 2).[42] *Gambol* and *Tc1* transposons are distinct clades that are separated by *mariner* (DD34D catalytic triad) and other families of the *IS630-Tc1-mariner* superfamily. Although *gambol* appears to be related to a few DD34E transposons from cyanobacteria and fungi, no *gambol*-like elements have been reported in any other insects or animals thus far. It is possible that *gambol* may be the founding member of a deep lineage of widely distributed DD34E transposons and additional *gambol*-like transposons from diverse organisms may be discovered as more genome sequences become available. Another family in the *IS630-Tc1-mariner* superfamily, DD37E, has also only been found in mosquitoes so far.[35] The mosquito-specific distribution of these unique transposon families is intriguing, although there is no obvious reason to suggest that mosquito genomes are particularly fertile grounds for unique families of DNA transposons.

LTR Retrotransposons: Diversity and Potentially Active Elements

A large number of LTR retrotransposons exist in the *An. gambiae* genome. For example, a detailed analysis uncovered 63 LTR retrotransposons in the *Ty3/gypsy* clade, although the copy number of these elements tends to be low (less than 30 copies).[41] This characteristic of low copy number but high diversity appears to be shared by elements in the *Ae. aegypti* genome (S. Li and Z. Tu, unpublished data). Approximately 75% of the *An. gambiae Ty3/gypsy* elements appear to have been at least recently active, on the basis of sequence analysis. There is experimental evidence of activity for three different LTR retrotransposons in *An. gambiae*. Biessmann et al[32] showed strong RNA expression of a LTR retrotransposon named *Moose* in the male and female gonads. Rohr et al[48] showed that *mtanga*, a *Ty1/copia* retrotransposon specific to the Y

Table 2. DNA-mediated transposable elements in mosquitoes

Superfamily/family	Defining features	Element	Organism
I. "Cut and paste" transposons			
P	8bp TSD[a], Conserved Transposase	*AgaP*[b] (8 subfamilies[39]), *AgPLS*[40]	*Anopheles gambiae*
		AqmP7,[39]	*Anopheles quadrimaculatus*
		AstP5[39]	*Anopheles stephensi*
		AalP9, AalP10[39]	*Anopheles albimanus*
hAT (hobo-Ac-Tam3)	8bp TSD, Conserved Transposase	*hAT-type*[36]	*Anopheles gambiae*
		Herves[59]	*Anopheles gambiae*
IS630-Tc1-mariner	TA TSD, DDE(D) catalytic triad		
Tc1	DD34E	*MsqTc3*[60]	*Aedes aegypti*
		Quetzal[61]	*Anopheles albimanus*
		Crusoe,[34] *DD34E, Tiang, Topi, Tsessebe*[36]	*Anopheles gambiae*
		Tango, Frisky, Sunny[42]	*Anopheles gambiae*
Gambol	DD34E	*Kiwi, Ozzie, Whistler, Mango, AgH1, Parker, Pags, Lovejoy, Watteau, Pi-Pi, Piper, Julo*[42]	*Anopheles gambiae*
mariner[c]	DD34D	*mariner*[36]	*Anopheles gambiae*
		Atmar-1, Atmar-2[44]	*Ochlerotatus atropalpus*
ITmD37D (maT)	DD37D	*ITmD37D*[36]	*Anopheles gambiae*
ITmD37E	DD37E	*A.aegypti.ITmD37E*[35]	*Aedes aegypti*
		A.gambiae.ITmD37E[35]	*Anopheles gambiae*
Pogo[d]	DDxD, long C-terminus	*Pogo-like*[36]	*Anopheles gambiae*
piggyBac	TTAA TSD, Conserved Transposase	*piggyBac*[36] (Sakar et al., unpublished)	*Anopheles gambiae*
		AgaPB1-10[43]	*Anopheles gambiae*
PIF-harbinger	3bp TSD, Conserved Transposase	*PIF/harbinger-like* (Biedler et al., unpublished)	*Anopheles gambiae*
Transib	5bp TSD, Conserved Transposase	*Transib1_AG*[15]	*Anopheles gambiae*
0 0r		*Transib1_AA, Transib2_AA, Transib3_AA, Transib4_AA, Transib5_AA*[7]	*Aedes aegypti*
		Ikirara[62]	*Anopheles gambiae*

continued on next page

Table 2. Continued

Superfamily/family	Defining features	Element	Organism
Not classified			
II. Rolling circle transposons			
Helitron	No TSD, Similarity to Helicase	*Helitron1_AG, Helitron2_AG* [15]	*Anopheles gambiae*
III. MITEs			
mTA	TA TSD	*Pony,* [63] *Wujin;* [31] *mTA_1-8* (Mao and Tu, unpublished)	*Aedes aegypti*
		TA-I-Ag, TA-II-Ag, TA-III-Ag, TA-IV-Ag, TA-V-Ag [64]	*Anopheles gambiae*
		Mikado, Milord, Mimo, Mirza, Nemo [18]	*Culex pipiens*
m3bp (Tourist-like MITEs)	3bp TSD	*m3bp_1-10* (Mao and Tu, unpublished)	*Aedes aegypti*
		Joey, TAA-I-Ag, TAA-II-Ag, [64] *m3bp_1-19* (Biedler et al., unpublished)	*Anopheles gambiae*
		A MITE in a NOS gene [65]	*Anophles stephensi*
m4bp	4bp (often TTAA) TSD	*Wukong, Wuneng,* [31] *m4bp_1-6* (Mao and Tu, unpublished)	*Aedes aegypti*
m8bp	8bp TSD	*m8bp_1-10* (Mao and Tu, unpublished)	*Aedes aegypti*
		8bp-I-Ag, [64] *Pegasus* [66]	*Anophles gambiae*
m9bp[e]	9bp TSD	*m9bp_1-2* (Mao and Tu, unpublished)	*Aedes aegypti*
Not classified		*Microuli (TTAA TSD, SIR,* [67])	*Aedes aegypti*
		Mint (CA TSD, SIR [18])	*Culex pipiens*

Notes

a. TSD, target site duplication.

b. A number of P-like elements were also reported in other members of the *An. gambiae* species complex.

c. *mariner* sequences, full-length or fragment, have been found in a wide range of insects. The DD34D mariners consist of 6 subfamilies; briggase, cecropia, elegans, irritans, mauritiana, mellifera. See Shao and Tu[35] for a re-classification of the *IS630-Tc1-mariner* superfamily. See Robertson[68] for reviews on *mariner* evolution.

d. The status of *Pogo* is not certain. It is either considered a member of the *IS630-Tc1-mariner* superfamily or a separate superfamily.[35, 68]

e. The existence of MITEs of 9 bp TSDs suggests that mosquitoes may contain mutator-like transposons, which so far are only found in plants and fungi.[27]

f. The citation is not always the original report of a TE. In some cases a review or an article describing the phylogenetic placement of the TE is cited instead.

g. There are PCR surveys of a few DNA-mediated TEs in mosquitoes that are not listed in the table.[47]

chromosome, was actively transcribed. One-LTR circles, which are potential replication intermediates, were detected of *mtanga*. Recently, a *Bel*-like element named *Belly*, was also shown to be potentially active in *An. gambiae* (M. Alvarez, M. Davenport, M. Jacobs-Lorena, and Z. Tu, unpublished data). Two short peptide sequences that are parts of the *Belly* protein were detected in the midgut. One-LTR circles of *Belly* were recovered from *An. gambiae* cells. The *An. gambiae* genome may be a rich source of active LTR retrotransposons that could be developed as genetic tools.

Non-LTR Retrotransposons: Unprecedented Diversity and Potentially Active Elements

An unprecedented diversity of non-LTR retrotransposons was discovered by analysis of the *An. gambiae* genome assembly.[37] More than 100 non-LTR retrotransposons were found by a reiterative and exhaustive search using the conserved reverse transcriptase (RT) domains of representatives of known non-LTRs as the starting queries. These non-LTRs range from a few to approximately 2000 copies and occupy at least 3% of the genome. *An. gambiae* non-LTRs represent eight of the 15 previously defined clades plus two novel clades Loner and Outcast. Five belong to the L1 clade, the first such elements in an invertebrate. Although most non-LTR elements appear to be inactive, there are approximately 20 elements that showed characteristics of recent activity. The incredible diversity and the maintenance of multiple recently active lineages within different clades are intriguing. Preliminary analysis of the *Ae. aegypti* genome assembly revealed more than 150 non-LTR retrotransposons that belong to 11 clades (J. Biedler and Z. Tu, unpublished data). This represents the greatest diversity in any reported genome. *Penelope*, an interesting group of retrotransposons that contain introns,[72] was also found in *Ae. aegypti* (Biedler and Tu, unpublished data). Some of the *Ae. aegypti* non-LTR retrotransposons are highly repetitive. For example, *JuanA*, an element in the Jockey clade, consists of more than 10,000 copies and occupies at least 3% of the entire genome. This high copy number is unexpected as previous estimation by Southern blot suggested approximately 200 copies.[30]

Gecko and MosquI

Gecko is a novel SINE discovered in *Ae. aegypti.*[57] It is shown that SINE transcripts are recognized by the retrotransposition machinery of its partner non-LTR retrotransposon through shared sequences or structures at their 3' termini.[11,12] The 3' regions of *gecko* and *MosquI* are similar in sequence and identical in secondary structure. *MosquI* is a potentially autonomous non-LTR retrotransposon in *Ae. aegypti* that is related to the *Drosophila I* factor.[53] It is possible that *MosquI* is the non-LTR retrotransposon "partner" of *gecko.* With respect to 3' termini, non-LTR retrotransposons are classified as poly(dA) elements such as human *L1* or elements with 3' tandem repeats such as the *Drosophila I* factor.[73] Some SINEs such as *Alu* end with poly(dA) while others including *Ae. aegypti Feilai* end with simple tandem repeats. What is unique about *gecko* is that it has different types of 3' termini, either poly(dA) or tandem repeats with variable sequences. Many of the *gecko* copies that end with different termini could not be differentiated when the 3' poly(dA) or repeats were excluded in phylogenetic and sequence analyses. Alteration of 3' termini during retrotransposition has been experimentally demonstrated using artificial constructs of the *Drosophila I* factor and the human Alu elements.[11,12] The analysis of *gecko* represents the first report of natural alterations between 3' tandem repeats and a poly(dA) tail in a SINE family. The 3' tandem repeats and poly(dA) tails may be generated by similar mechanisms during retrotransposition and *gecko* provides genomic and evolutionary support for the previously proposed slippage retrotransposition model.[57] The error-prone reverse transcription may generate the initial change in a 3' repeat unit in a retrotransposed copy of non-LTR element or SINE and the altered repeat unit could subsequently be replicated during slippage reverse transcription. Thus the distinction between poly(dA) and nonpoly(dA) elements may not be informative with regard to their origin and evolutionary relationship.

TE Display

TE display is a recently developed experimental method to detect TE insertions and it has the potential to significantly impact mosquito TE research.[74-76] It is a variation of Amplified Fragment Length Polymorphism (AFLP), the difference being that one primer is designed according to a TE sequence. First, genomic DNA is digested using a four-base restriction enzyme such as BfaI and then ligated to an adapter sequence. This is followed by two rounds of PCR with a radioactive-labeled nested primer specific for the TE sequence in the second round. Products are run on a sequencing gel and visualized by autoradiography. When conditions are optimized, each band represents a TE insertion site. Because TE display is a PCR-based approach, hundreds or more reactions can be run using genomic DNA isolated from individual mosquitoes. Therefore, many different TEs could be tested using the same DNA sample. TE display offers a higher degree of sensitivity and resolution than genomic Southern analysis. After TE display, one side of the TE insertion and its associated flanking sequence can be recovered because the band can be reamplified and sequenced. The insertion site may be precisely located if the genome sequence is available. TE display has been established with several families of MITEs,[76] SINEs (Y. Qi, A. della Torre, T. Scott, L. Harrington, G. Yan, and Z. Tu, unpublished data), and non-LTR retrotransposons[77] in *An. gambiae* and/or *Ae. aegypti*. The highly polymorphic nature of the insertions of *Herves*[59] and several non-LTR retrotransposons[77] was used as indirect evidence for current or recent transposition. One important advantage of TE display is the ability to rapidly scan a large number of TE insertions across a genome. Therefore TE display is useful to screen for polymorphic insertion sites and to detect new insertions that result from transposition.[78,79] For example, TE Display has been used to detect somatic cell transposition,[78] simply by looking for the presence of new bands that represent newly transposed copies of a TE. The same method may also be used to identify germline transposition by comparing TE display patterns of parent insects with the patterns of a large number of offspring.

Practical Applications

TEs and Genetic Manipulation of Mosquitoes

Active TEs can be used as vectors to insert genes of interest into the genome of a species to change its genetic makeup. TEs can also be used for gene trapping, enhancer trapping, and genome-wide insertional mutagenesis studies, which are powerful ways to investigate gene function and regulation on a genome scale.[80-82] A small number of DNA transposons such as *Hermes*, *MosI*, *minos*, and *piggyBac*, which are exogenous to mosquito species, show varied degrees of utility as transformation tools in mosquitoes.[26] Interestingly, *Hermes*, *MosI*, and *piggyBac* display noncanonical integration in the germline of *Ae. aegypti* and/or *Culex pipiens quinquefaciatus*.[83-86] For example, when a plasmid that contains an autonomous *Hermes* element was introduced into the *Ae. aegypti* germline, a complete copy of the plasmid was integrated as well as an adjoined rearranged partial copy.[85] Although somatic mobilization events were detected in the above transformants, no remobilization was detected in the germline. Improvement of current transformation vectors and development of new vectors are urgently needed and a better understanding of mosquito TEs will facilitate this effort. For example, a better understanding of mosquito TEs will help mitigate the interactions between exogenous vectors and endogenous transposons that share similar TIRs, which has been shown to be potentially problematic.[87,88] In addition, analysis of TEs in mosquito genomes will expand the pool of active TEs, which may be used to design a suite of tools with unique and useful features. As discussed earlier, *Herves*, a *hAT*-like transposon discovered in the *An. gambiae* genome was shown to be able to mobilize in *D. melanogaster*.[59] Therefore, *Herves* has the potential to be a new transformation vector. Because TEs has the potential to replicate in germline and rapidly spread in a population, scientists are exploring the possibility to use TEs to "drive" beneficial transgenes into insect populations.[89,90] This strategy is being investigated in the context of "driving" refractory genes into mosquito populations to control mosquito-borne infectious diseases.[91,92] A better understanding of the activity and regulation of endogenous TEs and their behavior in mosquito populations is critical to help achieve the sustained success of such sophisticated genetic approaches.

TE Insertions As Population Markers

TE display of a number of DNA and RNA-mediated elements showed high levels of insertion site polymorphism in *An. gambiae* and *Ae. aegypti* populations.[59,76,77] In most cases, TE display bands may only be used as dominant markers because individuals with insertions at both alleles cannot be distinguished from individuals that have TE insertion at only one of the two alleles. To mitigate this disadvantage, TE display can be used to identify and isolate TE insertion sites that are polymorphic in mosquito populations. Once such insertion sites are identified, sequences flanking a TE at a specific locus may be used as primers to amplify genomic DNA isolated from an individual. When the PCR products are run on an agarose gel, individuals with insertions at both alleles will show a single high molecular mass band while individuals with no insertions at either allele will give a single low molecular mass band. Individuals that have heterozygous alleles will display both bands. SINEs including the human *Alu* elements have been shown to be extremely powerful genetic markers using this locus-specific PCR method (e.g., refs. 93-96). SINE insertion markers have a few potential advantages over other interspersed markers such as microsatellite and Single Nucleotide Polymorphism.[96] The same SINE insertions at a specific locus are considered identical by descent and the ancestral state of SINE insertion polymorphism is known, which is the absence of a SINE. However, one has to be cautious because SINE excision and other rare confounding events do occur.[97] As shown in Table 1, several families of SINEs have been described in *Ae. aegypti, C. pipiens*, and *An. gambiae*. It is possible to efficiently screen for SINE insertion polymorphism in mosquitoes using TE display (Y. Qi, A. della Torre, T. Scott, L. Harrington, G. Yan, and Z. Tu, unpublished data).[76] TE display and TE-based locus-specific PCR are being used as genome-wide markers to investigate the genetic differences between *An. gambiae* populations, which is significant because genetic heterogeneity between mosquito populations can affect both the efficiency of disease transmission and the effectiveness of mosquito control measures.[98,99] There is genetic discontinuity in sympatric populations of *An. gambiae*.[100] Two "molecular forms", M and S, were defined on the basis of single nucleotide differences in rDNA sequences and this grouping was supported by the virtual absence of hybrid rDNA genotypes in nature. Microsatellite surveys suggest that the two forms are in the process of incipient speciation, with low levels of gene flow between forms homogenizing regions of the genome that are not directly involved in the speciation process. TE (or SINE) based markers may prove to be powerful new tools for population genomic analysis of the divergence between the M and S molecular forms of *An. gambiae*.

Future Perspectives

This is an exciting time for mosquito TE research. The diversity of recent discoveries highlighted in the previous sections represent only a modest beginning. The publication of the *An. gambiae* genome assembly[36] has propelled mosquito TE research into the genomic era. The nearly completed *Ae. aegypti* genome assembly has already become a tremendous resource for TE discovery and genomic and evolutionary comparisons of TEs in highly divergent mosquito species. In addition, sequencing of the *C. pipiens quinquefaciatus* genome, supported by the National Institute of Allergy and Infectious Diseases (NIAID), has already begun. The National Human Genome Research Institute (NHGRI) has approved the sequencing of the M and S forms of *An. gambiae*. A proposal is being considered which calls for sequencing of six additional species of Anopheline mosquitoes including two additional members of the *An. gambiae* complex (Nora Besansky at the University of Notre Dame, personal communication). Accompanying the genome sequencing of a growing list of mosquito species is the development of a number of powerful computational tools for the discovery and characterization of TEs on a whole-genome scale. These include several computational approaches that have been previously reviewed[27] and a few new tools that employ different strategies.[101,102] There are also efforts to develop systems biology methods to study mosquito TEs that are described below. The ongoing accumulation of genomic resources and recent advances in research methodology make it possible to address a number of exciting questions of fundamental and applied significance.

Impact on Gene Expression

There are a number of different ways by which TEs may affect host biology. It has been argued that TE insertions may interact with the regulatory elements of nearby genes and thus alter their expression.[5,103] The availability of a number of mosquito genomes offers the opportunity to systematically investigate the effect of TEs on nearby genes. Comparison of TE insertions in the genome assembly of M and S forms of *An. gambiae* and closely related species will reveal which fixed insertions are conserved. These TE insertions may affect the expression of nearby genes or provide new variation in transcripts/proteins, which may be examined experimentally. A large number of EST or cDNA sequences are often generated concurrently with genome sequencing projects. Indeed many TEs are found in EST or cDNA sequences in both *An. gambiae* and *Ae. aegypti* (Z. Tu, unpublished data), which begs the question of the functions of these TEs in expressed sequences.

Impact on Chromosomal Inversions

TE insertions may serve as substrates for recombination events that could lead to chromosomal inversions that are common in Anopheline mosquitoes.[104] A DNA TE named *Odysseus* was found adjacent to the distal breakpoint of a naturally occurring paracentric chromosomal inversion that is characteristic of *An. arabiensis*, one of the species in the *An. gambiae* complex.[105] TEs may indeed be important in generating chromosomal rearrangements in nature, which might lead to reproductive isolation and genetic changes that allow *An. gambiae* to exploit a range of ecological niches.[105] The availability of genome sequence data from the M and S forms of *An. gambiae* and related mosquitoes will be invaluable in studying the contribution of TEs on chromosomal inversions.

TEs and Genome Evolution

It has been proposed that variable TE abundance may account for the "C-value paradox", which reflects the discrepancy between genome size of an organism (as indicated by the C value) and its biological complexity.[106] It is not well-established to what extent and how TEs affect genome evolution. Genome size and organization varies significantly between different mosquitoes. The genome of *An. gambiae* is 270 Mbp in size and is organized in a pattern of "long period interspersion".[107] In contrast, the 1200 Mbp *Ae. aegypti* genome is organized in a pattern of mainly "short period interspersion"[108] (TIGR *Ae. aegypti* genome project website, http://www.tigr.org/tdb/e2k1/aabe/). The *C. pipiens* genome is approximately 540 Mbp.[107] Systematic analyses of the *An. gambiae, Ae. aegypti* and *C. pipiens* genomes will reveal how TEs contribute to the size and organization of divergent mosquito genomes. Intriguing intraspecific variation of genome size was reported during surveys of 47 populations of the Asian tiger mosquito, *Ae. albopictus*, ranging from 620 Mbp to 1660 Mbp.[107] This phenomenon offers a rare window into an early stage of genome size evolution. The availability of sequences of a large number of mosquito TEs will allow the use of systems biology tools such as TE microarray to study the contribution of TEs to such intraspecific variations using a Comparative Genomic Hybridization (CGH)[109] approach. A TE microarray including oligos of a few hundred mosquito TEs is being tested in our laboratory. The relative abundance of a large number of TEs may be compared between different populations/strains.

TEs in Mosquito Populations

One major challenge to our understanding of TE evolution is to determine how TEs are maintained and spread in natural populations. Moreover, TEs have been proposed as tools to drive the spread of beneficial genes through mosquito populations to control infectious diseases.[91,92] For such a sophisticated approach to work, it is critical to understand the natural population dynamics of TEs in their mosquito hosts. The combination of experimental methods such as TE display and locus-specific PCR, and the availability of a number of mosquito genomes such as the different forms of *An. gambiae*, may offer an excellent opportunity to investigate TE evolutionary dynamics in natural populations. Identification and isolation of insertions that are fixed in mosquito populations will be of special interest because of their evolutionary significance.

Evolutionary Origin of TE Diversity

High levels of diversity have been reported for both RNA-mediated and DNA-mediated TEs in *An. gambiae* and *Ae. aegypti*[37,41,42] (Z. Tu, unpublished data). The evolutionary process that generated this diversity is of great interest. It is possible that some TEs may be transmitted by a complex mix of horizontal transfer and vertical inheritance. It is also possible that some TEs may be under pressure to diversify within a genome to escape the fate of inactivation.[110] Parsing out the results of intragenomic diversification from those of horizontal transfer events has been difficult. However, comparative analysis of the TE landscape in sequenced genomes from a number of mosquitoe species having diverse evolutionary distance will provide opportunities to better understand the scenarios under which mosquito TEs evolve. As the number of characterized mosquito TEs increases, rare horizontal transfer events may also be uncovered through experimental surveys in which genomic DNA samples from a wide range of mosquito species are hybridized to all known mosquito TEs on a TE microarray.

Activity and Regulation

In the past, a majority of actively transposing TEs were discovered through fortuitous events such as the characterization of a mutant. As summarized in an earlier review,[27] it is now possible to use a systematic approach, combining computational and experimental methods, to uncover active TEs in a sequenced genome. A successful example of using such an approach to isolate active TEs in *An. gambiae* is provided by Arensburger and colleagues.[59] It is anticipated that more active TEs will be discovered this way as more genomes are sequenced. As described above, a TE array is being developed for comparative genomic hybridization. The same array is also used to detect transcription of a large number of TEs in mosquito cells. Transcription of a TE may indicate active transposition, although alternative explanations are possible. These new developments in conjunction with established methods such as plasmid mobility assays, germline transformation assays, and TE display will greatly facilitate the discovery of active TEs.

Another very interesting fundamental question about TEs is their regulation and interaction with host genome. There are a number of ways by which TEs are regulated in insects.[27] RNA interference (RNAi), a mechanism that confers post-transcriptional degradation of mRNA on the basis of homology to small fragments of double stranded RNA, has been implicated as a host defense mechanism against a broad spectrum of TEs in a nematode *Caenorhabditis elegans*.[111,112] RNAi has been shown to suppress viral infections in mosquitoes.[113] The above-mentioned TE array is also being used to compare the level of TE transcription in normal and RNAi-deficient mosquito cells. It is possible that release of suppression will increase transcription of potentially active TEs, leading to the identification of novel active TEs.

In summary, newly developed computational and experimental methods and the expanding genome revolution provide an exciting opportunity for the discovery and in-depth analysis of mosquito TEs, which will greatly improve our basic understanding of mosquito genetics and provide powerful tools for genetic manipulation of mosquitoes to help control mosquito-borne infectious diseases.

Acknowledgement

This work was supported by NIH grants AI42121, AI063252, and AI053203 and the Virginia Agricultural Experimental Station.

References

1. Craig N. Mobile DNA: An introduction. In: Craig N et al, eds. Mobile DNA II. American Society for Microbiology Press, 2002:3-11.
2. Doolittle WF, Sapienza C. Selfish genes, the phenotype paradigm and genome evolution. Nature 1980; 284(5757):601-3.
3. Orgel LE, Crick FH. Selfish DNA: The ultimate parasite. Nature 1980; 284(5757):604-7.
4. Brookfield JF. Molecular evolution. Retroposon revivals. Curr Biol 1995; 5(3):255-6.
5. Kidwell MG, Lisch DR. Transposable elements as sources of genomic variation. In: Craig N, Craigie R, Gellert M, Lambowitz A, eds. Mobile DNA II. American Society for Microbiology Press, 2002:59-90.
6. Agrawal A, Eastman QM, Schatz DG. Transposition mediated by RAG1 and RAG2 and its implications for the evolution of the immune system. Nature 1998; 394(6695):744-51.
7. Kapitonov VV, Jurka J. RAG1 core and V(D)J recombination signal sequences were derived from Transib transposons. PLoS Biol 2005; 3(6):e181.
8. Eickbush T, Malik H. Origins and evolution of retrotransposons. In: Craig N et al, eds. Mobile DNA II. American Society for Microbiology Press, 2002:1111-1144.
9. Luan DD, Korman MH, Jakubczak JL et al. Reverse transcription of R2Bm RNA is primed by a nick at the chromosomal target site: A mechanism for non-LTR retrotransposition. Cell 1993; 72(4):595-605.
10. Eickbush T. R2 and related site-specific nonlong terminal repeat retrotransposons. In: Craig N et al, eds. Mobile DNA II. Washington, DC: American Society for Microbiology Press, 2002:813-835.
11. Kajikawa M, Okada N. LINEs mobilize SINEs in the eel through a shared 3' sequence. Cell 2002; 111(3):433-44.
12. Dewannieux M, Esnault C, Heidmann T. LINE-mediated retrotransposition of marked Alu sequences. Nat Genet 2003; 35(1):41-8.
13. Okada N, Hamada M. The 3' ends of tRNA-derived SINEs originated from the 3' ends of LINEs: A new example from the bovine genome. J Mol Evol 1997; 44(Suppl 1):S52-6.
14. Ohshima K, Hamada M, Terai Y et al. The 3' ends of tRNA-derived short interspersed repetitive elements are derived from the 3' ends of long interspersed repetitive elements. Mol Cell Biol 1996; 16(7):3756-64.
15. Kapitonov VV, Jurka J. Molecular paleontology of transposable elements in the Drosophila melanogaster genome. Proc Natl Acad Sci USA 2003; 100(11):6569-74.
16. Kapitonov VV, Jurka J. Rolling-circle transposons in eukaryotes. Proc Natl Acad Sci USA 2001; 98(15):8714-9.
17. Finnegan DJ. Transposable elements. Curr Opin Genet Dev 1992; 2(6):861-7.
18. Feschotte C, Zhang X, Wessler SR. Miniature inverted-repeat transposable elements and their relationship to established DNA transposons. In: Craig N, Craigie R, Gellert M, Lambowitz A, eds. Mobile DNA II. American Society for Microbiology Press, 2002:1147-1158.
19. MacRae AF, Clegg MT. Evolution of Ac and Dsl elements in select grasses (Poaceae). Genetica 1992; 86(1-3):55-66.
20. Zhang X, Feschotte C, Zhang Q et al. P instability factor: An active maize transposon system associated with the amplification of Tourist-like MITEs and a new superfamily of transposases. Proc Natl Acad Sci USA 2001; 98(22):12572-7.
21. Feschotte C, Mouches C. Recent amplification of miniature inverted-repeat transposable elements in the vector mosquito Culex pipiens: Characterization of the Mimo family. Gene 2000; 250(1-2):109-16.
22. Izsvak Z, Ivics Z, Shimoda N et al. Short inverted-repeat transposable elements in teleost fish and implications for a mechanism of their amplification. J Mol Evol 1999; 48(1):13-21.
23. Le QH, Wright S, Yu Z et al. Transposon diversity in Arabidopsis thaliana. Proc Natl Acad Sci USA 2000; 97(13):7376-81.
24. Garcillan-Barcia MP, Bernales I, Mendiola MV et al. IS91 rolling-circle transposition. In: Craig N et al, eds. Mobile DNA II. Washington, DC: American Society for Microbiology Press, 2002:891-904.
25. Tu Z, Coates C. Mosquito transposable elements. Insect Biochem Mol Biol 2004; 34(7):631-44.
26. Handler AM, O'Brochta DA. Transposable elements for insect transformation. In: Gilbert L, Iatrou K, Gill S, eds. Comprehensive Molecular Insect Science. Vol 4. Oxford: Elsevier, 2005:437-474.

27. Tu Z. Insect transposable elements. In: Gilbert L, Iatrous K, Gill S, eds. Comprehensive Molecular Insect Science. Vol 4. Oxford: Elsevier, 2005:395-436.
28. Warren AM, Hughes MA, Crampton JM. Zebedee: A novel copia-Ty1 family of transposable elements in the genome of the medically important mosquito Aedes aegypti. Mol Gen Genet 1997; 254(5):505-13.
29. Besansky NJ. A retrotransposable element from the mosquito Anopheles gambiae. Mol Cell Biol 1990; 10(3):863-71.
30. Mouches C, Bensaadi N, Salvado JC. Characterization of a LINE retroposon dispersed in the genome of three nonsibling Aedes mosquito species. Gene 1992; 120(2):183-90.
31. Tu Z. Three novel families of miniature inverted-repeat transposable elements are associated with genes of the yellow fever mosquito, Aedes aegypti. Proc Natl Acad Sci USA 1997; 94(14):7475-80.
32. Biessmann H, Walter MF, Le D et al. Moose, a new family of LTR-retrotransposons in the mosquito Anopheles gambiae. Insect Mol Biol 1999; 8(2):201-12.
33. Feschotte C, Fourrier N, Desmons I et al. Birth of a retroposon: The twin SINE family from the vector mosquito Culex pipiens may have originated from a dimeric tRNA precursor. Mol Biol Evol 2001; 18(1):74-84.
34. Hill SR, Leung SS, Quercia NL et al. Ikirara insertions reveal five new Anopheles gambiae transposable elements in islands of repetitious sequence. J Mol Evol 2001; 52(3):215-31.
35. Shao H, Tu Z. Expanding the diversity of the IS630-Tc1-mariner superfamily: Discovery of a unique DD37E transposon and reclassification of the DD37D and DD39D transposons. Genetics 2001; 159(3):1103-15.
36. Holt RA, Subramanian GM, Halpern A et al. The genome sequence of the malaria mosquito Anopheles gambiae. Science 2002; 298(5591):129-49.
37. Biedler J, Tu Z. Non-LTR retrotransposons in the African malaria mosquito, Anopheles gambiae: Unprecedented diversity and evidence of recent activity. Mol Biol Evol 2003; 20(11):1811-25.
38. Quesneville H, Nouaud D, Anxolabehere D. Detection of new transposable element families in Drosophila melanogaster and Anopheles gambiae genomes. J Mol Evol 2003; 57(Suppl 1):S50-9.
39. Sarkar A, Sengupta R, Krzywinski J et al. P elements are found in the genomes of nematoceran insects of the genus Anopheles. Insect Biochem Mol Biol 2003; 33(4):381-7.
40. Oliveira de Carvalho M, Silva JC, Loreto EL. Analyses of P-like transposable element sequences from the genome of Anopheles gambiae. Insect Mol Biol 2004; 13(1):55-63.
41. Tubio JM, Naveira H, Costas J. Structural and evolutionary analyses of the Ty3/gypsy group of LTR retrotransposons in the genome of Anopheles gambiae. Mol Biol Evol 2005; 22(1):29-39.
42. Coy MR, Tu Z. Gambol and Tc1 are two distinct families of DD34E transposons: Analysis of the Anopheles gambiae genome expands the diversity of the IS630-Tc1-mariner superfamily. Insect Mol Biol 2005; 14(5):537-46.
43. Sarkar A, Sim C, Hong YS et al. Molecular evolutionary analysis of the widespread piggyBac transposon family and related "domesticated" sequences. Mol Genet Genomics 2003b; 270(2):173-80.
44. Zakharkin SO, Willis RL, Litvinova OV et al. Identification of two mariner-like elements in the genome of the mosquito Ochlerotatus atropalpus. Insect Biochem Mol Biol 2004; 34(4):377-86.
45. Rongnoparut P, Sirichotpakorn N, Rattanarithikul R et al. Sequence heterogeneity in copia-like retrotransposons in Anopheles (Diptera: Culicidae) in Thailand. J Med Entomol 1998; 35(5):771-7.
46. Cook JM, Martin J, Lewin A et al. Systematic screening of Anopheles mosquito genomes yields evidence for a major clade of Pao-like retrotransposons. Insect Mol Biol 2000; 9(1):109-17.
47. Imwong M, Sharpe RG, Kittayapong P et al. Distribution of the transposable element mariner in anopheline mosquitoes. Heredity 2000; 85(Pt 3):271-6.
48. Rohr CJ, Ranson H, Wang X et al. Structure and evolution of mtanga, a retrotransposon actively expressed on the Y chromosome of the African malaria vector Anopheles gambiae. Mol Biol Evol 2002; 19(2):149-62.
49. Eiglmeier K, Wincker P, Cattolico L et al. Comparative analysis of BAC and whole genome shotgun sequences from an Anopheles gambiae region related to Plasmodium encapsulation. Insect Biochem Mol Biol 2005; 35(8):799-814.
50. Besansky NJ, Paskewitz SM, Hamm DM et al. Distinct families of site-specific retrotransposons occupy identical positions in the rRNA genes of Anopheles gambiae. Mol Cell Biol 1992; 12(11):5102-10.
51. Kojima KK, Fujiwara H. Evolution of target specificity in R1 clade non-LTR retrotransposons. Mol Biol Evol 2003; 20(3):351-61.
52. Tu Z, Isoe J, Guzova JA. Structural, genomic, and phylogenetic analysis of Lian, a novel family of non-LTR retrotransposons in the yellow fever mosquito, Aedes aegypti. Mol Biol Evol 1998; 15(7):837-53.

53. Tu Z, Hill JJ. MosquI, a novel family of mosquito retrotransposons distantly related to the Drosophila I factors, may consist of elements of more than one origin. Mol Biol Evol 1999; 16(12):1675-86.
54. Besansky NJ, Bedell JA, Mukabayire O. Q: A new retrotransposon from the mosquito Anopheles gambiae. Insect Mol Biol 1994; 3(1):49-56.
55. Bensaadi-Merchermek N, Cagnon C, Desmons I et al. CM-gag, a transposable-like element reiterated in the genome of Culex pipiens mosquitoes, contains only a gag gene. Genetica 1997; 100(1-3):141-8.
56. Tu Z. Genomic and evolutionary analysis of Feilai, a diverse family of highly reiterated SINEs in the yellow fever mosquito, Aedes aegypti. Mol Biol Evol 1999; 16(6):760-72.
57. Tu Z, Li S, Mao C. The changing tails of a novel short interspersed element in Aedes aegypti: Genomic evidence for slippage retrotransposition and the relationship between 3' tandem repeats and the poly(dA) tail. Genetics 2004; 168(4):2037-47.
58. Tu Z. Maque, a family of extremely short interspersed repetitive elements: Characterization, possible mechanism of transposition, and evolutionary implications. Gene 2001; 263(1-2):247-53.
59. Arensburger P, Kim Y, Orsetti J et al. An active transposable element, Herves, from the African malaria mosquito Anopheles gambiae. Genetics 2005; 169(2):697-708.
60. Shao H, Qi Y, Tu Z. MsqTc3, a Tc3-like transposon in the yellow fever mosquito Aedes aegypti. Insect Mol Biol 2001; 10(5):421-5.
61. Ke Z, Grossman GL, Cornel AJ et al. Quetzal: A transposon of the Tc1 family in the mosquito Anopheles albimanus. Genetica 1996; 98(2):141-7.
62. Romans P, Bhattacharyya RK, Colavita A. Ikirara, a novel transposon family from the malaria vector mosquito, Anopheles gambiae. Insect Mol Biol 1998; 7(1):1-10.
63. Tu Z. Molecular and evolutionary analysis of two divergent subfamilies of a novel miniature inverted repeat transposable element in the yellow fever mosquito, Aedes aegypti. Mol Biol Evol 2000; 17(9):1313-25.
64. Tu Z. Eight novel families of miniature inverted repeat transposable elements in the African malaria mosquito, Anopheles gambiae. Proc Natl Acad Sci USA 2001; 98(4):1699-704.
65. Luckhart S, Rosenberg R. Gene structure and polymorphism of an invertebrate nitric oxide synthase gene. Gene 1999; 232(1):25-34.
66. Besansky NJ, Mukabayire O, Bedell JA et al. Pegasus, a small terminal inverted repeat transposable element found in the white gene of Anopheles gambiae. Genetica 1996; 98(2):119-29.
67. Tu Z, Orphanidis SP. Microuli, a family of miniature subterminal inverted-repeat transposable elements (MSITEs): Transposition without terminal inverted repeats. Mol Biol Evol 2001; 18(5):893-5.
68. Robertson H. Evolution of DNA transposons in eukaryotes. In: Craig N et al, eds. Mobile DNA II. American Society for Microbiology Press, 2002:1093-1110.
69. Henikoff S. Detection of Caenorhabditis transposon homologs in diverse organisms. New Biol 1992; 4(4):382-8.
70. Capy P, Vitalis R, Langin T et al. Relationships between transposable elements based upon the integrase-transposase domains: Is there a common ancestor? J Mol Evol 1996; 42(3):359-68.
71. Doak TG, Doerder FP, Jahn CL et al. A proposed superfamily of transposase genes: Transposon-like ciliated protozoa and a common "D35E" motif. Proc Natl Acad Sci USA 1994; 91(3):942-6.
72. Arkhipova IR, Pyatkov KI, Meselson M et al. Retroelements containing introns in diverse invertebrate taxa. Nat Genet 2003; 33(2):123-4.
73. Bucheton A, Busseau I, Teninges D. I elements in Drosophila melanogaster. In: Craig N et al, eds. Mobile DNA II. Washington, DC: American Society for Microbiology Press, 2002:796-812.
74. Van den Broeck D, Maes T, Sauer M et al. Transposon Display identifies individual transposable elements in high copy number lines. The Plant Journal 1998; 13:121-129.
75. Casa AM, Brouwer C, Nagel A et al. Inaugural article: The MITE family heartbreaker (Hbr): Molecular markers in maize. Proc Natl Acad Sci USA 2000; 97(18):10083-9.
76. Biedler J, Qi Y, Holligan D et al. Transposable element (TE) display and rapid detection of TE insertion polymorphism in the Anopheles gambiae species complex. Insect Mol Biol 2003; 12(3):211-6.
77. Biedler J. Ph. D. theses. Biochemistry. Virginia Polytechnic Institute, 2005.
78. De Keukeleire P, Maes T, Sauer M et al. Analysis by Transposon Display of the behavior of the dTph1 element family during ontogeny and inbreeding of Petunia hybrida. Mol Genet Genomics 2001; 265(1):72-81.
79. Jiang N, Bao Z, Zhang X et al. An active DNA transposon family in rice. Nature 2003; 421(6919):163-7.

80. Horn C, Offen N, Nystedt S et al. piggyBac-based insertional mutagenesis and enhancer detection as a tool for functional insect genomics. Genetics 2003; 163(2):647-61.
81. Spradling AC, Stern D, Beaton A et al. The Berkeley Drosophila Genome Project gene disruption project: Single P-element insertions mutating 25% of vital Drosophila genes. Genetics 1999; 153(1):135-77.
82. Klinakis AG, Loukeris TG, Pavlopoulos A et al. Mobility assays confirm the broad host-range activity of the transposable element and validate new transformation tools. Insect Mol Biol 2000; 9(3):269-75.
83. Coates CJ, Jasinskiene N, Morgan D et al. Purified mariner (Mos1) transposase catalyzes the integration of marked elements into the germ-line of the yellow fever mosquito, Aedes aegypti. Insect Biochem Mol Biol 2000; 30(11):1003-8.
84. Wilson R, Orsetti J, Klocko AD et al. Post-integration behavior of a Mos1 mariner gene vector in Aedes aegypti. Insect Biochem Mol Biol 2003; 33(9):853-63.
85. O'Brochta DA, Sethuraman N, Wilson R et al. Gene vector and transposable element behavior in mosquitoes. J Exp Biol 2003; 206(Pt 21):3823-34.
86. Adelman ZN, Jasinskiene N, Vally KJ et al. Formation and loss of large, unstable tandem arrays of the piggyBac transposable element in the yellow fever mosquito, Aedes aegypti. Transgenic Res 2004; 13(5):411-25.
87. Sundararajan P, Atkinson PW, O'Brochta DA. Transposable element interactions in insects: Crossmobilization of hobo and Hermes. Insect Mol Biol 1999; 8(3):359-68.
88. Jasinskiene N, Coates CJ, James AA. Structure of hermes integrations in the germline of the yellow fever mosquito, Aedes aegypti. Insect Mol Biol 2000; 9(1):11-8.
89. Ribeiro JM, Kidwell MG. Transposable elements as population drive mechanisms: Specification of critical parameter values. J Med Entomol 1994; 31(1):10-6.
90. Engels WR. Invasions of P elements. Genetics 1997; 145(1):11-5.
91. Ashburner M, Hoy MA, Peloquin JJ. Prospects for the genetic transformation of arthropods. Insect Mol Biol 1998; 7(3):201-13.
92. Alphey L, Beard CB, Billingsley P et al. Malaria control with genetically manipulated insect vectors. Science 2002; 298(5591):119-21.
93. Batzer MA, Deininger PL. Alu repeats and human genomic diversity. Nat Rev Genet 2002; 3(5):370-9.
94. Batzer MA, Stoneking M, Alegria-Hartman M et al. African origin of human-specific polymorphic Alu insertions. Proc Natl Acad Sci USA 1994; 91(25):12288-92.
95. Salem AH, Kilroy GE, Watkins WS et al. Recently integrated Alu elements and human genomic diversity. Mol Biol Evol 2003; 20(8):1349-61.
96. Stoneking M, Fontius JJ, Clifford SL et al. Alu insertion polymorphisms and human evolution: Evidence for a larger population size in Africa. Genome Res 1997; 7(11):1061-71.
97. Medstrand P, van de Lagemaat LN, Mager DL. Retroelement distributions in the human genome: Variations associated with age and proximity to genes. Genome Res 2002; 12(10):1483-95.
98. Coluzzi M, Sabatini A, della Torre A et al. A polytene chromosome analysis of the Anopheles gambiae species complex. Science 2002; 298(5597):1415-8.
99. della Torre A, Costantini C, Besansky NJ et al. Speciation within Anopheles gambiae—the glass is half full. Science 2002; 298(5591):115-7.
100. della Torre A, Tu Z, Petrarca V. On the distribution and genetic differentiation of Anopheles gambiae s.s. molecular forms. Insect Biochem Mol Biol 2005; 35(7):755-69.
101. Edgar RC, Myers EW. PILER: Identification and classification of genomic repeats. Bioinformatics 2005; 21(Suppl 1):i152-i158.
102. Price AL, Jones NC, Pevzner PA. De novo identification of repeat families in large genomes. Bioinformatics 2005; 21(Suppl 1):i351-i358.
103. Muotri AR, Chu VT, Marchetto MC et al. Somatic mosaicism in neuronal precursor cells mediated by L1 retrotransposition. Nature 2005; 435(7044):903-10.
104. Sharakhov IV, Serazin AC, Grushko OG et al. Inversions and gene order shuffling in Anopheles gambiae and A. funestus. Science 2002; 298(5591):182-5.
105. Mathiopoulos KD, della Torre A, Santolamazza F et al. Are chromosomal inversions induced by transposable elements? A paradigm from the malaria mosquito Anopheles gambiae. Parassitologia 1999; 41(1-3):119-23.
106. Kidwell MG. Transposable elements and the evolution of genome size in eukaryotes. Genetica 2002; 115(1):49-63.
107. Rai KS, Black WCT. Mosquito genomes: Structure, organization, and evolution. Adv Genet 1999; 41:1-33.

108. Warren AM, Crampton JM. The Aedes aegypti genome: Complexity and organization. Genet Res 1991; 58(3):225-32.
109. Buckley PG, Mantripragada KK, Piotrowski A et al. Copy-number polymorphisms: Mining the tip of an iceberg. Trends Genet 2005; 21(6):315-7.
110. Lampe DJ, Walden KK, Robertson HM. Loss of transposase-DNA interaction may underlie the divergence of mariner family transposable elements and the ability of more than one mariner to occupy the same genome. Mol Biol Evol 2001; 18(6):954-61.
111. Ketting RF, Haverkamp TH, van Luenen HG et al. Mut-7 of C. elegans, required for transposon silencing and RNA interference, is a homolog of Werner syndrome helicase and RNaseD. Cell 1999; 99(2):133-41.
112. Tabara H, Sarkissian M, Kelly WG et al. The rde-1 gene, RNA interference, and transposon silencing in C. elegans. Cell 1999; 99(2):123-32.
113. Keene KM, Foy BD, Sanchez-Vargas I et al. RNA interference acts as a natural antiviral response to O'nyong-nyong virus (Alphavirus; Togaviridae) infection of Anopheles gambiae. Proc Natl Acad Sci USA 2004; 101(49):17240-5.

Chapter 3

Retrotransposons in the Genomes of the Digenean Parasitic Trematodes, *Clonorchis sinensis* and *Paragonimus westermani*

Young-An Bae and Yoon Kong*

Abstract

Considerable fractions of almost all eukaryotic genomes are composed of a variety of transposable elements (TEs). Members of the phylum Platyhelminthes are also thought to contain diverse TEs which comprise up to 40% of their genomes. A total of 29 retrotransposons, positioning into one non-long terminal-repeat (LTR) family (6 elements in *CR1*) and 3 LTR families (5 elements in *Xena* and *Bel*, and 13 elements in *Gypsy*) have been isolated from the genomes of the digenean trematodes, *Clonorchis sinensis* and *Paragonimus westermani*. *CsRn1* of *C. sinensis* and *PwRn1* of *P. westermani* are novel retrotransposons from full units for which the molecular structures have been established in detail. These elements were evenly distributed throughout the entirety of the respective genomes, with high copy numbers. They exhibited the characteristic structural features of LTR retrotransposons belonging to the *Gypsy* family and encoded each for a single long open reading frame. Phylogenetic analysis revealed that the *CsRn1* and *PwRn1* elements formed a novel, tightly-conserved clade, which has been suggested to have evolved uniquely in the metazoan genomes. Significant intergenomic variations induced by the heterogeneous integration of active copies of *CsRn1* were discovered among individual *C. sinensis*. These data suggested that diverse retrotransposon families are present in the lower animal taxa, and that some of these elements comprise important intermediate forms emerging during the course of evolution of the LTR retrotransposons. Retrotransposons in the trematodes might influence the remodeling of their host genomes.

Introduction

Retrotransposons comprise a common class of genetic elements which are mobile by virtue of the reverse transcription of RNA intermediates. They are widely distributed in almost all eukaryotic genomes. Retrotransposons are generally categorized into two large groups, long terminal-repeat (LTR) and non-LTR retrotransposons, according to their overall structures. Recent phylogenetic analyses based on the amino acid sequences of the Pol proteins have demonstrated that each of the two groups is composed of several distinct clades, members of which are believed to be closely related with one another, with regard to molecular evolution.[1-3] Retrotransposons are thought to have evolved from a common ancestor,[4] and it is also supposed that a certain lineage of non-LTR retrotransposons is responsible for the evolution of the LTR retrotransposons.[3]

*Corresponding Author: Yoon Kong—Department of Molecular Parasitology and Center for Molecular Medicine, Sungkyunkwan University and Samsung Biomedical Research Institute, Suwon 440-746, Korea. Email: ykong@med.skku.ac.kr

Mobile Genetic Elements in Metazoan Parasites, edited by Paul J. Brindley.

The expansion of retrotransposons might result in damaging effects to the host genome, including insertional mutagenesis and chromosomal translocations, both of which might be involved in the inactivation of adjacent gene(s), and the disruption of chromosomal integrity. However, it has also been proposed that most classes of retrotransposons are likely to coevolve with their hosts.[5-8] The integration of these elements into new loci within genomes might be associated with genomic variations ranging from simple sequence polymorphisms to large-scale structural alterations in the chromosomes. These alterations might be followed by the changes in the expression profiles of functional genes. Therefore, retrotransposons are believed to be major agents in evolution, giving rise to phenotypic variants, and, in the long-term, driving speciation.[9-10] These elements have recently been recognized as informative biomarkers, to compensate for the weakness of the currently-available molecular phylogenetic studies, most of which had been based on the sequence data of ribosomal and mitochondrial DNAs.[11-13]

The genomes of platyhelminths are highly repetitive,[14,15] which suggested that these organisms might be valuable as biological models for the study of the evolutionary roles of retrotransposons. However, relatively little studies have focused on the retrotransposons in the Platyhelminthes except for those of the human blood flukes of the genus *Schistosoma*.[16] Recent investigations have been taken to characterize the retrotransposons residing in the genomes of the human-infecting digenean parasitic trematodes, *Clonorchis sinensis* (oriental liver fluke) and *Paragonimus westermani* (lung fluke). This chapter provides information regarding the structures and genomic implications of the retrotransposons which were identified in these trematodes.

Biology and Genomes of *C. sinensis* and *P. westermani*

The phylum Platyhelminthes is a large and diverse group of organisms and, historically, has included four classes, namely Turbellaria, Monogenea, Trematoda, and Cestoda. Except for the free-living turbellarians, most of platyhelminths are parasitic, thriving in, or on, diverse species of vertebrate and invertebrate hosts. *C. sinensis* and *P. westermani* are taxonomically classified into the class Trematoda, of which members require at least two intermediate hosts to complete their life cycles as digenetic parasites. Miracidia in eggs of these parasites develop into cercaria in the first intermediate hosts (snails), during which the parasites are expanded by asexual reproduction. The cercariae, thereafter, become metacercariae in the second intermediate hosts (freshwater fishes for *C. sinensis*; crustaceans for *P. westermani*). The metacercariae can infect the final definitive hosts (various species of omnivorous mammals including humans) when the definitive host ingests the raw or undercooked second intermediate hosts. After excysting in human duodenum, the juvenile worms migrate to their predilection organs such as hepatic biliary ducts (*C. sinensis*) and lung (*P. westermani*), and mature into adults. The adults produce eggs in the definitive hosts via sexual reproduction. *C. sinensis* is the only species belonging to the genus *Clonorchis*, although the closest neighbors of the organism can be found in the genus *Opisthorchis*.[17] *C. sinensis* and *O. viverrini* cause hepatobiliary symptoms which are characterized by periductal fibrosis combined with ductal dilatation and adenomatous metaplasia. Patients heavily infected with *Clonorchis* or *Opistorchis* parasites present with jaundice with an abnormal liver function, and even abdominal distension by ascites. Chronic infections with these parasites might be associated with cholangiocarcinoma.[18] The genus *Paragonimus* includes more than 40 species, of which *P. westermani* represents the proto-type species and is widely distributed in Asia.[19] *P. westermani* mainly causes benign granuloma in the lung, but sometimes migrates aberrantly and invokes granulomatous lesions in the affected sites including the brain.[20]

The genomes of platyhelminths are shown to be highly variable in size ranging from 270 to 14,800 Mb and the percent repetitive sequences are comparable with those of higher vertebrates.[14] However, exact genomic sizes and contents of repetitive sequences of the parasitic helminths largely remained to be elusive. Such information is largely restricted to a very few species including *Mesostoma ehrenbergii*,[21] *Schistosoma mansoni*,[22] and *Spirometra mansonoides*.[23]

The large fractions of repetitive sequences found in these parasites suggest that numerous transposable elements (TEs) might exist in the genomes of platyhelminths.[16] The numbers of chromosome pairs of platyhelminths are also diverse and those of *C. sinensis* and *P. westermani* have been determined to be 28[24] and 11,[25] respectively. *C. sinensis* contains normal diploid genomes, whereas the populations of *P. westermani* display distinct contents in their genomes; i.e., di-, tri-, and tetrapolyploidy according to their geographical distributions.[19,25] The hybridization process is known to be a potent causal factor resulting in elevation of the mobile potentials of various retrotransposons.[26,27] Therefore, *P. westermani* could be considered as an important model system to investigate the expansion histories of retrotransposons and resulting effects on genomic reconstruction after hybridization.

Retrotransposon Families in *C. sinensis* and *P. westermani*

Several studies have concentrated on the isolation and characterization of TEs in *Schistosoma* spp. through which a number of Class I TEs, including the LTR retrotransposons, non-LTR retrotransposons, and SINEs have presently been described.[16,28] However, TEs of the other platyhelminths mostly remain unidentified and only small pieces of information are available on these elements from *C. sinensis* and *P. westermani.*

The partial sequences of various retrotransposons from the genomes of *C. sinensis* and *P. westermani* were retrieved and analyzed by a degenerate PCR method, which has proven suitable for the amplification of LTR elements. Sequence homology patterns and phylogenetic analysis revealed that these elements were highly divergent and were separately located in the *Xena*, *Bel*, and *Gypsy* families of LTR retrotransposons (Table 1). A retrotransposon-related sequence belonging to the *Gypsy* family has been isolated from the *C. sinensis* genome by arbitrarily primed PCR (AP-PCR).[29] *CR1*-like non-LTR retrotransposons were also fortuitously identified during the degenerate PCR screening (Table 1).

Conversely, members of the *Copia* family were not identified in these trematode genomes. The main difference between *Copia* and the other families involves the position of the integrase (IN) domain. In the *Copia* family, it is located upstream of the reverse transcriptase (RT)/RNase H (RH) domains, but in the other families, it is located downstream of the RT/RH. The IN domain is the largest one encoded by *pol* genes of retrotransposons. *Copia* members are ubiquitously distributed throughout the evolutionary taxa,[30] and have been previously isolated from the genome of *Anopheles gambiae* with primers identical to those used in this screening.[31] Therefore, the failure with regard to the retrieval of *Copia* sequences might be resultant from difficulties in the amplification of relatively large fragments under the PCR conditions used. *DIRS*-like elements have also not been identified, to date, in the genomes of liver and lung flukes. Since the distribution of the *DIRS* family across evolutionary taxa is still a matter of much debate, it remains impossible to conclude whether or not the trematode genomes truly contain *DIRS* family members. Despite these unanswered questions, it is apparent that diverse LTR and non-LTR retrotransposon families are present in the genomes of *C. sinensis* and *P. westermani.*

Structures of Retrotransposons

A total of 29 retrotransposons, including LTR and non-LTR elements, have been described in the genomes of *C. sinensis* and *P. westermani.* These elements revealed characteristic amino acid conservation in their open reading frames (ORFs), according to the families to which they belonged. However, overall structural features of these elements could not be determined in detail mainly due to the incomplete nature of their sequence data. The full unit and structural feature of the retrotransposon, designated as *CsRn1* has been comprehensively characterized only in *C. sinensis* (Fig. 1).[29]

The *CsRn1* element (GenBank accession no. AY013569), with an overall size of 5,026 bp, was flanked by 471-bp LTRs. As with most LTR retrotransposons, the LTR sequences contained short inverted repeats of 3 bp at both ends, which begun with TG and terminated

Table 1. *Retrotransposon families in the genomes of* Clonorchis sinensis *and* Paragonimus westermani

Family	Name	Organism	Accession No.	Reference
Non-LTR retrotransposon				
CR1	*CsCR1*	*C. sinensis*	CW848111	
	CsCR2	*C. sinensis*	CW848114	
	CsCR3	*C. sinensis*	CW848115	
	CsCR4	*C. sinensis*	CW848116	
	CsCR5	*C. sinensis*	CW848118	
	CsR1	*C. sinensis*	CW848113	
LTR retrotransposon				
Xena	*CsRn5*	*C. sinensis*	CW848109	
	PwRn6	*P. westermani*	CW848126	33
	PwRn7	*P. westermani*	CW848130	33
	PwRn8	*P. westermani*	CW848128	33
	PwRn9	*P. westermani*	CW848123	33
Bel	*CsRn3*	*C. sinensis*	CW848107	
	CsRn7	*C. sinensis*	CW848112	
	CsRn8	*C. sinensis*	CW848120	
	PwRn12	*P. westermani*	CW848129	33
	PwRn13	*P. westermani*	CW848132	33
Gypsy	*CsRn1*	*C. sinensis*	AY013569	29
	CsRn2	*C. sinensis*	CW848106	
	CsRn4	*C. sinensis*	CW848108	
	CsRn6	*C. sinensis*	CW848110	
	CsRn9	*C. sinensis*	CW848117	
	CsRn10	*C. sinensis*	CW848119	
	PwRn1	*P. westermani*	AY237162	
	PwRn2	*P. westermani*	CW848127	33
	PwRn3	*P. westermani*	CW848121	33
	PwRn4	*P. westermani*	CW848125	33
	PwRn5	*P. westermani*	CW848122	33
	PwRn10	*P. westermani*	CW848124	33
	PwRn11	*P. westermani*	CW848131	33

with CA (5'-TGT...ACA-3'). This element was shown to use the $tRNA^{Trp}$ molecule as a primer for the synthesis of the first cDNA strand during expansion, as the 12-bp sequence located adjacent to the 5'-LTR is complementary to the nucleotides at the 3'-end of bovine and chicken $tRNA^{Trp}$. The putative poly-purine tract (PPT) for the synthesis of the second cDNA strand (5'-GGGGGAGTAG-3') was located in the direct upstream region of the 3'-LTR.

LTR retrotransposons possess *gag* and *pol* genes which are required for their replication. In some cases, they were also found to have *env* gene. The internal region of *CsRn1* encoded for a single long ORF composed of 1,304 amino acids. The deduced ORF included well-conserved functional protein domains, in the order of protease (PR), RT, RH, and IN. The ORF was found to contain no Env-like protein. Amino acid motifs detected in the active sites of these protein domains were also highly conserved in the ORF of *CsRn1* (DTG in PR; YLDD in RT; DAS in RH; HHCC and DDE motifs in IN). In general, low levels of sequence identity are observed among the Gag proteins of LTR retrotransposons, except for a copy of the $CX_2CX_4HX_4C$ (CCHC) motif, which is similar to those recognized in the

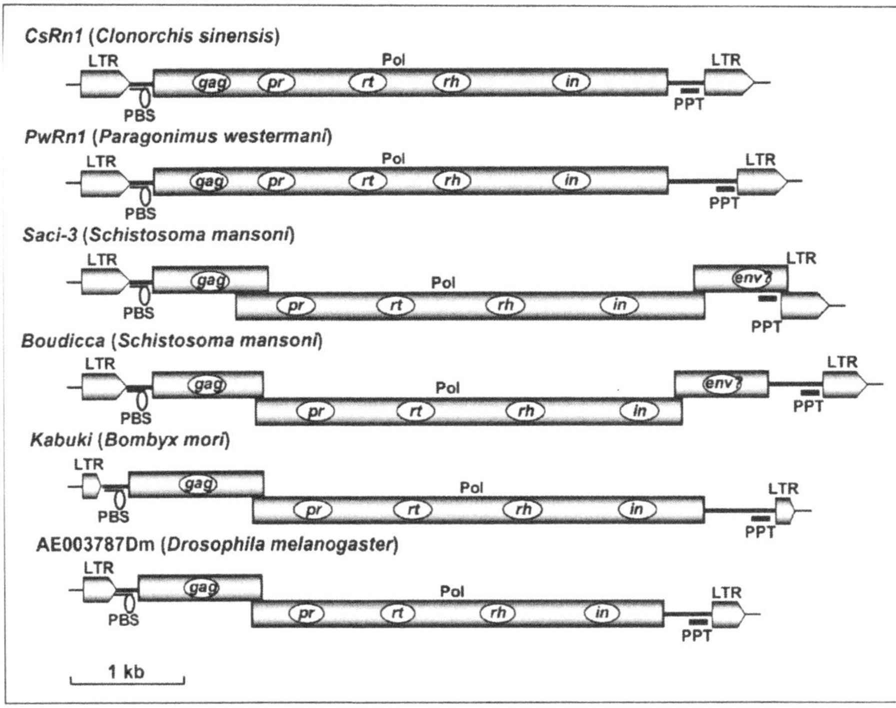

Figure 1. Schematic representation of *CsRn1* of *Clonorchis sinensis* and *PwRn1* of *Paragonimus westermani*, and related retrotransposons. The overall structures of these elements were compared with those of *Kabuki* from *Bombyx mori* (AB032718), an undescribed retrotransposon from *Drosophila melanogaster* (AE003787), and *Saci-3* (BK004070) and *Boudicca* (BK004066) from *Schistosoma mansoni*. Abbreviations used are: LTR = long terminal repeat; ORF = open reading frame; TSD = target site duplication; PBS = primer binding site; PPT = polypurine tract; pr = protease; rt = reverse transcriptase; rh = RNase H; and in = integrase.

basic nucleocapsid portion of retroviral Gag polyproteins. However, this conventional motif was not detected in the upstream region of *CsRn1* PR. The region was found to contain a domain which exhibited significant homology with the putative Gag proteins of *Kabuki* of *Bombyx mori*, an undescribed LTR retrotransposon of *Drosophila melanogaster* on AE003787, and *Saci-3* and *Boudicca* of *S. mansoni*.[28,29] These elements possessed the well-conserved $CX_2HX_9CX_3C$ (CHCC) sequence motif rather than the conventional CCHC motif. In addition, a homology search using the TBLASTN algorithm revealed that retrotransposons with the unconventional CHCC motif in their ORFs were ubiquitous in the genome of *Anopheles gambiae* (e.g., retrotransposons on AAAB01008980, AAAB01008859, and AAAB01008968). Considering the amino acid conservation and enzymatic domain order (PR-RT-RH-IN) in the deduced ORF, *CsRn1* could be a member of the *Gypsy*-like LTR retrotransposons. It seemed likely that the clade including these elements was huge, comprising numerous retrotransposons in platyhelminth and insect genomes.

The full unit of a retrotransposon designated as *PwRn1* (GenBank accession no. AY237162) was recently isolated from the genome of *P. westermani* via degenerate PCR techniques. The overall structure and coding profile of *PwRn1* are similar to those of *CsRn1*, except that *PwRn1* is much larger, with an overall size of 5,400 bp (Fig. 1). In addition to the well-conserved functional motifs in the Pol protein, the unique CHCC motif was also found in a putative Gag protein of the element (unpublished observation).

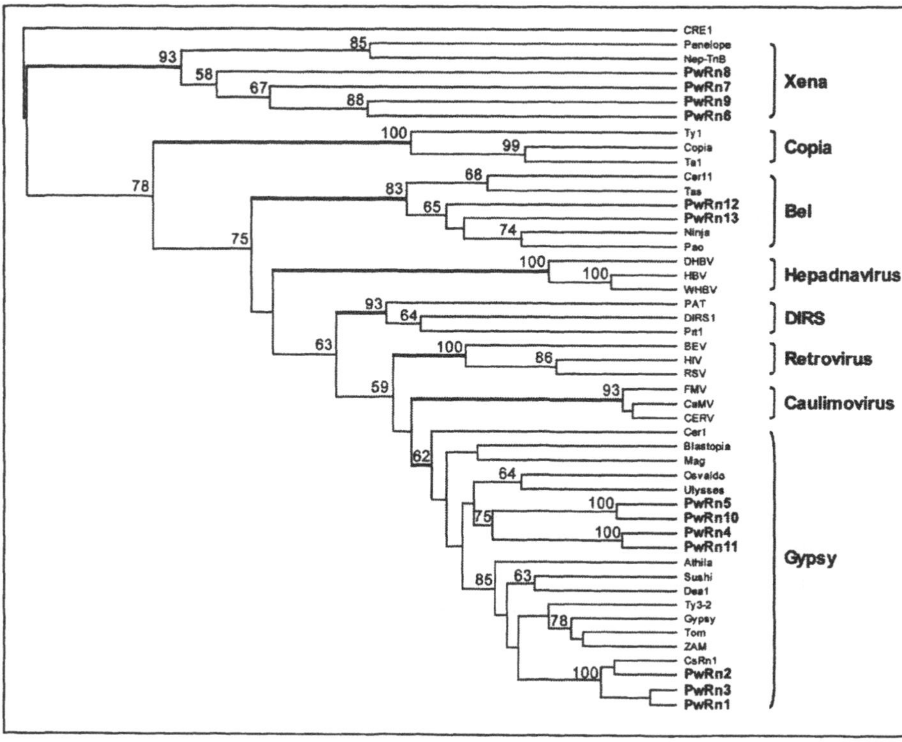

Figure 2. The phylogenetic relationships between retrotransposons of *Paragonimus westermani* and the known long terminal-repeat (LTR) containing genetic elements. The tree was constructed based on a sequence alignment with the amino acids of reverse transcriptase domains, from I to IV, employing the maximum parsimony algorithm implemented in PHYLIP. The statistical significance of each branching node is presented as bootstrap values of 100 replicates. The tree is rooted with the non-LTR retrotransposon *CRE1*. Descriptions of the elements in bold-letters represent those of *P. westermani*.

Evolutionary Insights

It is important to understand the genetic events occurring in the initial stage of molecular evolution. According to recent cumulative data on retrotransposons, the evolutionary origins of LTR-containing genetic elements, including retroviruses, have been extensively analyzed on the basis of the amino acid sequences of the RTs and RHs.[3] In relation to this model, a previous study suggested that the members of the *Xena* family are the most ancient of these elements, and diverged during an early stage of LTR retrotransposon evolution.[32] *Xena*-like elements have been detected in a wide range of taxa including insects, deuterostomes and echinoderms, and have been shown to share certain common structural features.[32] However, information regarding *Xena*-like elements is still fairly sparse, as only a few studies have examined the retrotransposons which belong to this family. Several *Xena*-like elements have been discovered in the genomes of *C. sinensis* and *P. westermani*. Members of the *Bel* family, regarded as an important intermediate form, are also widely distributed in these parasitic genomes (Table 1 and Fig. 2).

The *Gypsy* family is believed to have emerged most recently and has been divided into eight well-conserved clades.[2,3] A phylogenetic analysis, including the trematode retrotransposons indicated the presence of previously unidentified additional clades of the

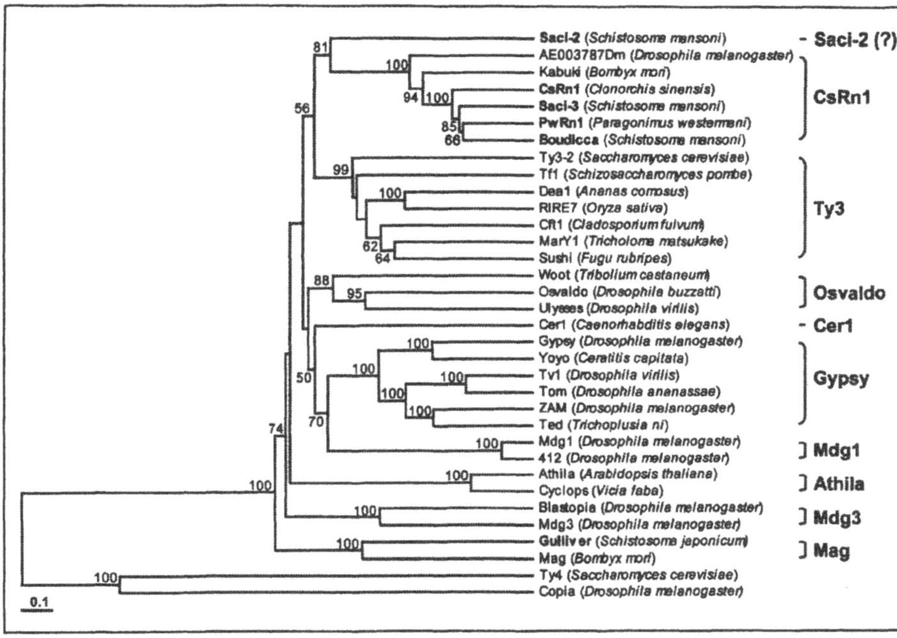

Figure 3. Evolutionary positions of *CsRn1* and *PwRn1* among long terminal-repeat retrotransposons (LTR) belonging to the *Gypsy* family. The phylogenetic analysis was based on an alignment of amino acid sequences from reverse transcriptase to the central subdomain of integrase (DDE motif). The tree was constructed using the UPGMA algorithm using PHYLIP, and rooted with the *Ty4* and *Copia*, which are phylogenetically distinct groups of the LTR retrotransposons. The bootstrap values are shown as numbers at the branching points. The nomenclatures for nine, well-conserved clades are also presented. Retrotransposons identified from the Platyhelminthes are indicated by boldface font.

Gypsy family. Four elements of *P. westermani*, which were classified into the *Gypsy* family (*PwRn4*, *-5*, *-10*, and *-11*), have been shown to form one or two distinct clade(s), which cluster differently from known clade members of the family. *CsRn1* and *CsRn4* of *C. sinensis* exhibited significant homology with *Kabuki* of *B. mori*. These elements were clustered into another tightly-conserved clade with *PwRn1*, *PwRn2*, and *PwRn3* of *P. westermani*. This clade, designated *CsRn1* clade, also contained *Kabuki*, *Saci-3*, and *Boudicca* as well as the undescribed retrotransposon of *D. melanogaster* (Fig. 3). As members of a freshly defined clade, these elements shared a number of common features including a unique CHCC Gag motif, target site duplication (TSD) of 4 bp, sequence conservation in the primer-binding site (PBS) and PPT, and the functional protein domains. The *CsRn1* clade includes retrotransposons which have been identified largely in trematode and insect genomes. This result strongly suggested that members of this clade have evolved exclusively in the metazoan genomes, and that they constituted an important variant form which emerged during the evolutionary course of the *Gypsy*-like LTR retrotransposons.

Mobile Potentials and Distribution Patterns

Retrotransposons replicate their progeny copies via the reverse transcription of mRNA transcripts. Therefore, the initial step in the expansion of a retrotransposon is the transcription of the element itself, using host RNA polymerase. The mobile potential of a given element can be predicted by determining the presence or absence of its mRNA transcripts.

The mRNA transcripts of *CsRn1* were detected in the *C. sinensis* adult worms, indicating that the element was currently preserving the mobile potentials. In accord with this finding, uncorrupted ORFs were maintained in certain copies of *CsRn1*, and a certain degree of genomic variations induced by the heterogeneous integration of active *CsRn1* copies was identified among individuals in the *C. sinensis* population.[29] In *P. westermani*, mRNA transcripts of several retrotransposons were also recognized.[33] Taken together, it is strongly suggested that retrotransposons actively replicate their progenies, thus affecting the remodeling of their host genomes in these trematode parasites.

The heterogeneous integration of retrotransposons might induce a significant polymorphism among individual genomes, and thus reflect the recent spread of retrotransposons that maintain the mobile potentials. The polymorphic distribution of *CsRn1* was also observed in the *C. sinensis* population, as detected by interretrotransposon amplified polymorphism (IRAP) analysis (Fig. 4A). If a copy of a retrotransposon has integrated into a genomic locus in an early evolutionary era, the retrotransposon-occupying locus might be fixed in all haploid genomes. Conversely, if the retrotransposon copy has been replicated in recent times, the genomic locus with the element might be detected in a certain fraction of haploid genomes. The evolutionary course of multiple *CsRn1* copies has been examined by the analysis of their LTR and RT sequences.[34] When insertional polymorphisms of these copies were traced by retrotransposon-based insertional polymorphism (RBIP) analysis, the degree to which polymorphism occurred varied substantially; younger elements were heterogeneous and older elements were homogeneous, at the intra- and interpopulation levels (Fig. 4B,C).

The *CsRn1* element had a copy number of approximately 100 per haploid genome, and was distributed evenly throughout the entirety of the *C. sinensis* genome. The distribution pattern of *PwRn1* in the *P. westermani* genome was quite similar to that observed with *CsRn1*, and existed with a copy number of more than 1,000 per haploid genome. In general, the target sites are randomly selected during the retrotransposition process, although some elements exhibit a certain degree of preference for restricted genomic loci in their target site selection, such as *R1* and *R2* for the ribosomal DNA locus and *TART* and *HeT-A* for the chromosomal ends, in the genome of *D. melanogaster*. Analyses of the flanking regions of the *CsRn1* copies revealed that this element tends to integrate into agenic chromosomal regions.[34]

Genomic Implications

There have been several suggestions, none of which are mutually exclusive, addressing the preferential distributions of retrotransposons in heterochromatic or agenic regions of the host genomes.[35-37] Based on the hypothesis that average insertions tend to be neutral (at best) or deleterious to the host, retrotransposons would be better tolerated in the heterochromatin, due to the paucity of the functional genes in this region. Reduced recombination during meiosis is also relevant for the accumulation of the elements in this region.[38] Alternatively, hotspots for the insertion of retrotransposons might exist in the heterochromatin, especially considering the highly repetitive nature of the sequences normally present in this region. All of these proposed mechanisms are consistent with a great deal of collected data, albeit a certain degree of controversial data has also been reported. The actual mechanisms underlying the accumulation of retrotransposons in the heterochromatin remained largely uncertain.

Recently expanded *CsRn1* copies were largely found in agenic parts of the host chromosomes.[34] Since no information regarding the chromosomal structures of *C. sinensis* is presently available, it is hard to conclude whether these regions actually represent heterochromatins, or whether they are parts of the euchromatin, in which the density of functional genes happens to be relatively low. The clustered distributions of the TEs within the heterochromatin have been described in *D. melanogaster*.[39,40] When analyzing the nucleotide sequences flanking *CsRn1*, a considerable portion of the sequences matched those of the *pol* or *gag* of retrotransposons.[34] This result implied that *CsRn1* is normally coclustered with other retrotransposons, in what is probably heterochromatin.

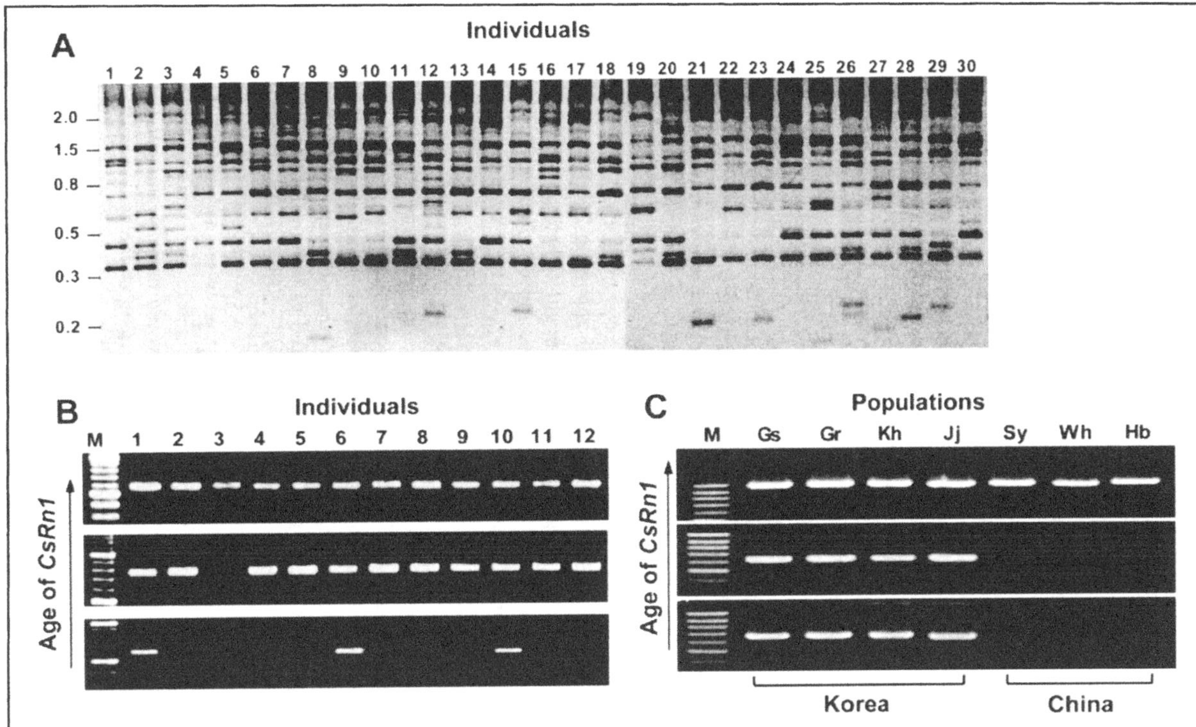

Figure 4. Intergenomic distribution patterns of *CsRn1*. A) Banding patterns of interretrotransposon amplified polymorphism (IRAP) analysis, using individual *Clonorchis sinensis* genomes. B) Retrotransposon-based insertional polymorphism (RBIP) analysis of *CsRn1* using individual *C. sinensis* genomes. Numerals in A (1-30) and B (1-12) represent individuals of *C. sinensis,* from which genomic DNAs were extracted. C) RBIP analysis of *CsRn1* among *C. sinensis* populations. The evolutionary ages of *CsRn1* copies were estimated by analyzing the degrees of sequence divergence in their long terminal-repeats (LTRs).[34] For comparing *CsRn1*-occupying loci at the interpopulation level, *C. sinensis* specimens were collected from endemic areas of Korea (Gs= Gokseung; Gr= Gurye; Kh= Kimhae; Jj= Jinju) and China (Sy= Shenyang; Wh=Wuhan; Hb= Harbin).

TEs located in the heterochromatin have been less highlighted especially in terms of genetics and cytology, and thus their roles in this region have been remained unclear. Some evidence of the stability and structural rearrangement of TEs have been suggested, namely that the inserted elements in this region fall easily into progressive decay, leading to the ultimate extinction of the elements.[41] However, many researches have indicated that TEs play significant roles in the maintenance and evolution of their host genomes. In *D. melanogaster*, the erosion of telomeres due to incomplete replication is protected by the continuous retrotransposition of *TART* and *HeT-A* at the ends of the chromosomes.[42,43] It has also been hypothesized that TEs are involved in the evolution of tandemly repeated DNAs in the heterochromatin.[44]

TEs can also influence the expression profiles of the host functional genes in the euchromatin. Human *LINE-1* is known to be a major genetic element which mediates the inactivation of numerous functional genes located in a homologous X chromosome in females.[45] In the mouse genome, active *intracisternal A particle* (*IAP*) retrotransposons have been shown to be capable of modifying the expression pattern of a functional gene in a heritable manner.[46] The modifications of gene expression profiles are beneficial to the host under certain circumstances, but are damaging in other cases. In addition to the changes in the expression of functional genes, certain retrotransposons are more directly related to the survivals of their hosts, providing specific proteins originated directly from the elements themselves.[47,48]

A recent investigation has established that approximately half of the nucleotide sequences of the analyzed *CsRn1*-occupying loci exhibited no homology with any hitherto-known genes.[34] This observation might result, in part, from the scarcity of information regarding the genes of *C. sinensis* and, in part, from the small numbers of loci analyzed. The effects of *CsRn1* on the expression of functional genes and resulting phenotypic variations cannot be properly estimated at this moment. *CsRn1* induced a considerable degree of genomic polymorphism among individual *C. sinensis* genome (Fig. 4A). *CsRn1* is also isolated in the genome of *Opisthorchis viverrini*, the closest neighbor of *C. sinensis*, with similar copy numbers. The unique integration patterns of the element have been detected between these cousin species (unpublished observation). The heterogeneous integration of *CsRn1* might induce phenotypic variations among *C. sinensis* individuals and/or between *C. sinensis* and *O. viverrini*, by affecting the expression patterns of the nearby functional genes.

Concluding Remarks

In recent years, numerous retrotransposons have been isolated from the genomes of the parasitic trematodes, *C. sinensis* and *P. westermani*. These elements are positioned separately with regard to the diverse retrotransposon families, emerging at different times during the evolution of various reverse-transcribing elements. Some of these elements demonstrate unique clustering patterns upon phylogenetic analysis, compared to those exhibited by the previously described retrotransposons. This fact may reflect the presence of undetected retrotransposon clades in the trematode genomes, which are thought to function as important evolutionary intermediates. The evolutionary courses associated with these retrotransposons are probably highly sophisticated.[3] Therefore, the isolation and characterization of retrotransposons belonging to novel clades would broaden our current status of knowledge and provide more comprehensive understanding of the evolutionary history of these elements. It is quite apparent that there are more diverse retrotransposons in the platyhelminth genomes, although their copy numbers tend not to be quite so high as those of the higher vertebrates. In this respect, platyhelminths constitute useful model systems for the study of retrotransposons.

Extensive studies have been conducted on the development of genetic markers or statistical methods for phylogenetic analysis among organisms at both intra- and interspecific levels. For this purpose, allozymes, restriction fragment length polymorphisms (RFLP), and the nucleotide sequences of ribosomal and mitochondrial DNAs have been mainly targeted. Despite their general usefulness, these biomarkers occasionally show limitations with regard to the establishment of precise evolutionary relationships among the populations of a species, due to the restricted number of possible allelic forms and/or the extremely low level of sequence divergence. Retrotransposons are apparently advantageous, in that these elements make it possible to trace the flow paths of haplotypes in a species, as they become irreversibly integrated into new genomic loci in the course of evolutionary time.[49] The integration patterns exhibited by *CsRn1* and other retrotransposons, when evaluated by IRAP and RBIP analyses, can be used as informative data sets in phylogenetic studies among trematode populations. At present, the genomic implications of retrotransposons residing in trematode genomes cannot be properly evaluated. This is mainly due to the limited available information regarding the structures and sequences of host genomes and retrotransposons. Accumulation of more comprehensive data regarding the trematode genomes, and the complete structures and distribution patterns of retrotransposons other than *CsRn1* and *PwRn1*, would provide new insights into the evolutionary history of the retrotransposons and their roles in the host genomes.

References

1. Malik HS, Burke WD, Eickbush TH. The age and evolution of non-LTR retrotransposable elements. Mol Biol Evol 1999; 16:793-805.
2. Malik HS, Eickbush TH. Modular evolution of the integrase domain in the Ty3/Gypsy class of LTR retrotransposons. J Virol 1999; 73:5186-5190.
3. Malik HS, Eickbush TH. Phylogenetic analysis of ribonuclease H domains suggests a late, chimeric origin of LTR retrotransposable elements and retroviruses. Genome Res 2001; 11:1187-1197.
4. Xiong Y, Eickbush TH. Origin and evolution of retroelements based upon their reverse transcriptase sequences. EMBO J 1990; 9:3353-3362.
5. Finnegan DJ. Eukaryotic transposable elements and genome evolution. Trends Genet 1989; 5:103-107.
6. Csink AK, McDonald JF. Analysis of Copia sequence variation within and between Drosophila species. Mol Biol Evol 1995; 12:83-93.
7. Kidwell MG, Lisch D. Transposable elements as sources of variation in animals and plants. Proc Natl Acad Sci USA 1997; 94:7704-7711.
8. Fedoroff N. Transposons and genome evolution in plants. Proc Natl Acad Sci USA 2000; 97:7002-7007.
9. Lönnig WE, Saedler H. Plant transposons: Contributors to evolution? Gene 1997; 205:245-253.
10. Long AD, Lyman RF, Morgan AH et al. Both naturally occurring insertions of transposable elements and intermediate frequency polymorphisms at the achaete-scute complex are associated with variation in bristle number in Drosophila melanogaster. Genetics 2000; 154:1255-1269.
11. Takahashi K, Terai Y, Nishida M et al. A novel family of short interspersed repetitive elements (SINEs) from cichlids: The patterns of insertion of SINEs at orthologous loci support the proposed monophyly of four major groups of cichlid fishes in lake Tanganyika. Mol Biol Evol 1998; 15:391-407.
12. Verneau O, Catzeflis F, Furano AV. Determining and dating recent rodent speciation events by using L1 (LINE-1) retrotransposons. Proc Natl Acad Sci USA 1998; 95:11284-11289.
13. Nikaido M, Rooney AP, Okada N. Phylogenetic relationships among cetartiodactyls based on insertions of short and long interspersed elements: Hippopotamuses are the closest extant relatives of whales. Proc Natl Acad Sci USA 1999; 96:10261-10266.
14. Regev A, Lamb MJ, Jablonka E. The role of DNA methylation in invertebrates: Developmental regulation or genome defense? Mol Biol Evol 1998; 15:880-891.
15. Johnston DA, Blaxter ML, Degrave WM et al. Genomics and the biology of parasites. BioEssays 1999; 21:131-147.
16. Brindley PJ, Laha T, McManus PD et al. Mobile genetic elements colonizing the genomes of metazoan parasites. Trends Parasitol 2003; 19:79-87.
17. King S, Scholz T. Trematodes of the family Opisthorchiidae: A minireview. Korean J Parasitol 2001; 39:209-221.
18. Choi BI, Han JK, Hong ST et al. Clonorchiasis and cholangiocarcinoma: Etiologic relationship and imaging diagnosis. Clin Microbial Rev 2004; 17:540-552.

19. Blair D, Xu ZB, Agatsuma T. Paragonimiasis and the genus Paragonimus. Adv Parasitol 1999; 42:113-222.
20. Im JG, Kong Y, Shin YM et al. Pulmonary paragonimiasis: Clinical and experimental studies. RadioGraphics 1993; 13:575-586.
21. Hebert PDN, Beaton MJ. Breeding system and genome size of the rhabdocoel turbellarian Mesostoma ehrenbergii. Genome 1990; 33:719-724.
22. Oliveira G, Rodrigues NB, Romanha AJ et al. Genome and genomics of Schistosomes. Can J Zool 2004; 82:375-390.
23. Cox GS, Phares CK, Schmidt RA. Molecular characterization of the Spirometra mansonoides genome: Renaturation kinetics, methylation, and hybridization to human cDNA probes. Biochim Biophys Acta 1990; 1049:134-144.
24. Park GM, Im KI, Huh S et al. Chromosomes of the liver fluke, Clonorchis sinensis. Korean J Parasitol 2000; 38:201-206.
25. Park GM, Lee KJ, Im KI et al. Occurrence of a diploid type and a new first intermediate host of a human lung fluke, Paragonimus westermani, in Korea. Exp Parasitol 2001; 99:206-212.
26. Evgen'ev MB, Zelentsova H, Shostak N et al. Penelope, a new family of transposable elements and its possible role in hybrid dysgenesis in Drosophila virilis. Proc Natl Acad Sci USA 1997; 94:196-201.
27. Labrador M, Farre M, Utzet F et al. Interspecific hybridization increase transposition rates of Osvaldo. Mol Biol Evol 1999; 16:931-937.
28. DeMarco R, Kowaltowski AT, Machado AA et al. Saci-1, -2, and -3 and Perere, four novel retrotransposons with high transcriptional activities from the human parasite Schistosoma mansoni. J Virol 2004; 78:2967-2978.
29. Bae YA, Moon SY, Kong Y et al. CsRn1, a novel active retrotransposon in a parasitic trematode, Clonorchis sinensis, discloses a new phylogenetic clade of Ty3/gypsy-like LTR retrotransposons. Mol Biol Evol 2001; 18:1474-1483.
30. Peterson-Burch BD, Voytas DF. Genes of the Pseudoviridae (Ty1/copia retrotransposons). Mol Biol Evol 2002; 19:1832-1845.
31. Cook JM, Martin J, Lewin A et al. Systematic screening of Anopheles mosquito genomes yields evidence for a major clade of Pao-like retrotransposons. Insect Mol Biol 2000; 9:109-117.
32. Dalle Nogare DE, Clark MS, Elgar G et al. Xena, a full-length basal retroelement from tetraodontid fish. Mol Biol Evol 2002; 19:247-255.
33. Bae YA, Kong Y. Divergent long-terminal-repeat retrotransposon families in the genome of Paragonimus westermani. Korean J Parasitol 2003; 41:221-231.
34. Bae YA, Kong Y. Evolutionary course of CsRn1 long-terminal-repeat retrotransposon and its heterogeneous integrations into the genome of the liver fluke, Clonorchis sinensis. Korean J Parasitol 2003; 41:209-219.
35. Ganguly R, Swanson KD, Ray K et al. A BamHI repeat is predominantly associated with the degenerating neo-Y chromosome of Drosophila miranda but absent in the Drosophila melanogaster genome. Proc Natl Acad Sci USA 1992; 89:1340-1344.
36. Charlesworth B, Jarne P, Assimacopoulos S. The distribution of transposable elements within and between chromosomes in a population of Drosophila melanogaster. III. Element abundances in heterochromatin. Genet Res 1994; 64:183-197.
37. Junakovic N, Terrinoni A, Di Franco C et al. Accumulation of transposable elements in the heterochromatin and on the Y chromosome of Drosophila simulans and Drosophila melanogaster. J Mol Evol 1998; 46:661-668.
38. Sniegowski PD, Charlesworth B. Transposable element numbers in cosmopolitan inversions from natural population of Drosophila melanogaster. Genetics 1994; 137:815-827.
39. Carmena M, Gonzalez C. Transposable elements map in a conserved pattern of distribution extending from β-heterochromatin to centromeres in Drosophila melanogaster. Chromosoma 1995; 103:676-684.
40. Pimpinelli S, Berloco M, Fanti L et al. Transposable elements are stable structural components of Drosophila melanogaster heterochromatin. Proc Natl Acad Sci USA 1995; 92:3804-3808.
41. Vaury C, Bucheton A, Pélisson A. The β-heterochromatin sequences flanking the I elements are themselves defective transposable elements. Chromosoma 1989; 98:215-224.
42. Biessmann H, Kasravi B, Jakes K et al. The genomic organization of HeT-A retroposons in Drosophila melanogaster. Chromosoma 1993; 102:297-305.
43. Levis RW, Ganesan R, Houtchens K et al. Transposons in place of telomeric repeats at a Drosophila telomere. Cell 1993; 75:1083-1093.
44. Thompson-Stewart D, Karpen GH, Spradling AC. A transposable element can drive the concerted evolution of tandemly repetitious DNA. Proc Natl Acad Sci USA 1994; 91:9042-9046.

45. Bailey JA, Carrel L, Chakravarti A et al. Molecular evidence for a relationship between LINE-1 elements and X chromosome inactivation: The Lyon repeat hypothesis. Proc Natl Acad Sci USA 2000; 97:6634-6639.
46. Morgan HD, Sutherland HG, Martin DI et al. Epigenetic inheritance at the agouti locus in the mouse. Nat Genet 1999; 23:314-318.
47. Villarreal LP. On viruses, sex, and motherhood. J Virol 1997; 71:859-865.
48. Mi S, Lee X, Li X et al. Syncytin is a captive retroviral envelope protein involved in human placental morphogenesis. Nature 2000; 403:785-789.
49. Okada N. SINEs. Curr Opin Genet Dev 1991; 1:498-504.

Chapter 4

Endogenous Retrotransposon Sequences of the *Schistosoma mansoni* Intermediate Snail Host, *Biomphalaria glabrata*

Matty Knight,* Joanna Bridger, Wannaporn Ittiprasert, Edwin Odoemelam, Julio Masabanda, André Miller and Nithya Raghavan

Abstract

Schistosomiasis remains a major public health concern in 75 countries. Disease prevalence and transmission is intimately associated with the presence of freshwater snails that serve as intermediate hosts for the larval stage of the parasite. The discovery of retrotransposons and transposons in the parasite genome (reviewed in other chapters) offers some hope that Mobile Genetic Elements (MGEs) developed as delivery vectors may provide alternate tools that will help to disrupt disease transmission. In this chapter, we will discuss what is known about the existence of MGEs in the snail host. In particular, we will focus on the discovery and characterization of the first full-length non-Long Terminal Repeat (LTR)-retrotransposon identified from mollusks (named *nimbus, BgI*) from the snail *Biomphalaria glabrata* that transmits the digenean parasite *Schistosoma mansoni* in the Western Hemisphere. The *nimbus* (*BgI*) retrotransposon was identified from a Bacterial Artificial Chromosome (BAC) clone isolated by cross-hybridization with probes from a previously characterized Reverse Transcriptase (RT) encoded by a 2.0 Kb *Bam*H1 genomic fragment (BGR2) and an Expressed Sequence Tag (EST) *Bg37*, isolated from a hemocyte cDNA library of the snail. The *nimbus* (*BgI*) retrotransposon was found to belong to the I-clade and thus displayed the hallmark features such as the Open Reading Frame (ORF) 1, ORF 2, putative Target Site Duplications (TSDs), and terminal tandem repeats characteristic of retrotransposons of this clade. Several unique characteristics, however, especially in the flanking 5' and 3' Noncoding regions (NCRs), were detected in the snail element. The possible horizontal transfer of *nimbus* (*BgI*) in *S. mansoni* is also discussed, highlighting the need for further investigations into the role of MGEs within the context of host/parasite relationships.

Introduction

Sequences representing Mobile Genetic Elements (MGEs) are ubiquitous in the genomes of both plants and animals. It is believed that their ability to colonize and disperse widely throughout these genomes may have been a major driving force in evolution towards the establishment of genetic diversity and speciation.[1] Although the role of MGEs in causing deleterious mutations is well-known very few studies have been conducted on their role (if any) in the complex dynamics of host/parasite relationships or in the coevolution of pathogens and their

*Corresponding Author: Matty Knight—Biomedical Research Institute, 12111 Parklawn Drive, Rockville, MD 20852, USA. Email: mknight@afbr-bri.com.

Mobile Genetic Elements in Metazoan Parasites, edited by Paul J. Brindley.

hosts. Given that the biology of schistosome parasites obligates that they switch between an invertebrate (snail) and a vertebrate host to complete their life cycles, one may argue that this intimate coexistence, dating back millions of years, may have provided an opportunity for the lateral transfer of MGEs between these parasites and their hosts, even perhaps helping to shape the compatible outcomes of these ancient relationships.

In the case of the metazoan human parasites *Schistosoma mansoni* and *S. japonicum*, it is now clear that sequences representing several different MGEs exist in the genomes of these organisms (see other chapters in this book on schistosome retrotransposons and transposons). In this chapter, we will describe the discovery of sequences that represent MGEs from *Biomphalaria glabrata*, a major intermediate snail host of *S. mansoni*. It is hoped that characterization of MGEs in this snail host will help to address questions relating to their possible involvement in the molecular coevolution in host/parasite interactions. By characterizing these snail MGEs, it is also possible that vectors based on these sequences can be developed for the genetic transformation of both snails and schistosomes. With the success of germ line transformation in insects[2-3] and mollusks[4] using mobile genetic elements (MGEs), it is conceivable that this strategy can be extended to *B. glabrata* if MGEs can be identified and characterized in a way that will make the genetic manipulation of this snail a practical possibility. It is also anticipated that in conjunction with other well-established control measures (e.g., mass chemotherapy and the use of molluscides), development of novel tools that will also target the intermediate snail stage of the parasite's life cycle becomes important.[5] It is conceivable that genetically modified snails, carrying genes that confer resistance to infection, will one day help reduce the spread of schistosomiasis, a disease that has been difficult to control.[6]

Very little is known about the full repertoire of MGEs in the snail genome. However, from earlier experiments that involved the selection by self-fertilization of these hermaphroditic snails for various physical characteristics,[7] it is plausible that 'jumping genes' could be responsible for some of the non-Mendelian expression of certain abnormal traits (e.g., bulbous tentacles, multiple eyes, everted preputium, conjoined embryos) frequently observed in these snails.[8] Though the genetics governing the random appearance of these traits among inbred snails was shown to be complex, the possible involvement of MGEs as causing some of these abnormalities cannot be ruled out. It was not until the 1990s that studies towards a molecular understanding of *B. glabrata* in relation to trematode infections (schistosomes and echinostomes) were undertaken. Since then, however, considerable progress has been made towards the identification of genes underlying the snail host/parasite relationship.[9-13] In addition, the entire nucleotide sequence of the 930 Mb genome of *B. glabrata* is now being sequenced,[14] the results of which should provide us with more definitive information on the organization, number and distribution of MGEs in the snail genome.

Identification of a Non-LTR Retrotransposon-Related Sequence in *B. glabrata*: BGR2

Evidence of the existence of a sequence in the *B. glabrata* genome showing homology to known MGEs was first reported in 1992.[15] This sequence, named BGR2, was found to occur as a repetitive 1.95 Kb *Bam* HI fragment in the snail genome. The entire ORF of BGR2 showed a significant match with the RT domains of the human non-LTR retrotransposon, *L1*,[16] and the Drosophila I factor.[17] These non-LTR retrotransposable elements are the most abundant,[18-19] comprise 17% of the human genome and play a powerful role in modulating gene expression. The discovery of the BGR2 sequence in *B. glabrata* was made while studies involving the use of Southern hybridizations were being conducted to identify Restriction Fragment Length Polymorphisms (RFLPs) between genomes of parasite resistant and susceptible snails. Figure 1A shows the hybridization of the cloned BGR2 fragment (^{32}P labeled) to digested (*Bam* HI and *Pst* I) resistant (BS-90; lanes 1, 4) and susceptible (M-line; lanes 2, 5) snail DNAs, respectively and *S. mansoni* DNA (lane 3, 6). With the subsequent identification of an active cDNA transcript *Bg37* homologous to BGR2, the same blot in Figure 1A was

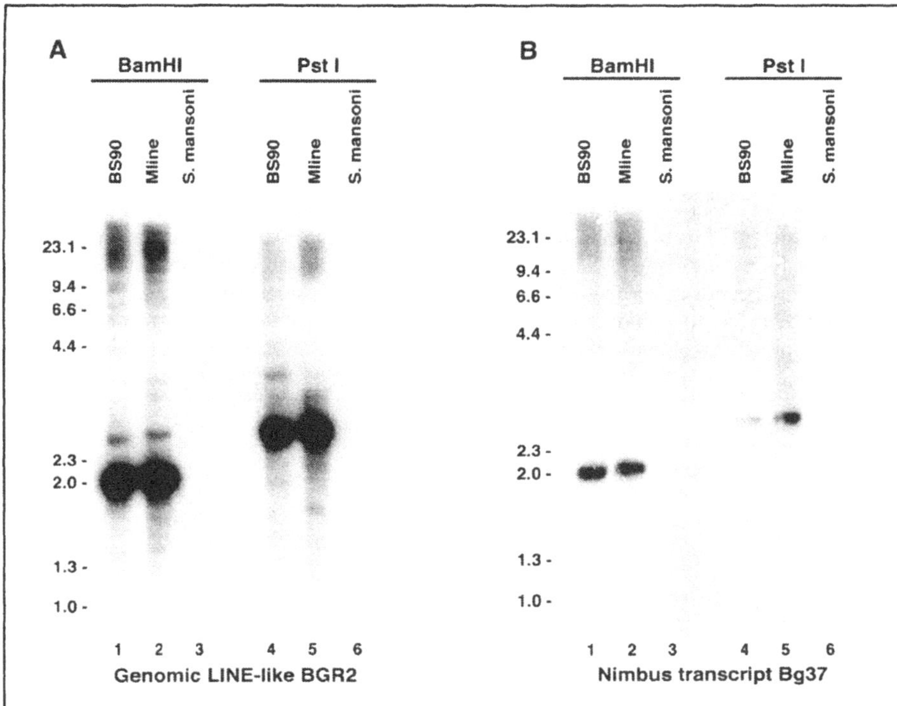

Figure 1. Southern hybridization of *B. glabrata*. 5 µg of genomic DNA from *B. glabrata* resistant (BS-90) and susceptible (M-line) snails and the parasite (*S. mansoni*) DNA were digested with either *Bam* HI or *Pst* I and probed using either the ^{32}P-labeled 1.95 Kb insert from the genomic clone BGR2 (GenBank X60372 and PIR PC1123) (Panel A) or the ^{32}P-labeled 476 bp *nimbus* (*BgI*) transcript *Bg37* (GenBank EF413179 and EST GenBank Acc. No. AW740305) (Panel B). Reprinted from: Raghavan N et al. Int J Parasitol 2007; doi:10.1016/j.ipara. 2007.04.002;[20] with permission from Elsevier, ©2007.

probed with this transcript. Figure 1B shows the hybridization of the labeled *Bg37* cDNA to digested (*Bam* HI and *Pst* I) resistant (BS-90; lanes 1, 4) and susceptible (M-line; lanes 2, 5) snail DNAs, respectively and *S. mansoni* DNA (lane 3, 6). Both the genomic BGR2 and the cDNA *Bg37* probes hybridized to a 2.0 Kb *Bam* HI and a 3.5 Kb *Pst* I fragment respectively. Though Southern blot and restriction enzyme analysis consistently failed to detect the BGR2 sequence in the parasite, results showed that the organization of the repetitive sequence was highly conserved in *B. glabrata* snails. Furthermore, by PCR analysis of DNA using primers designed from the BGR2 sequence, it is now clear that this MGE is widely distributed among snails of the neotropics and the Old World and is, therefore, not limited to laboratory maintained snails in which this element was first detected.[20]

Identification of ESTs Encoding MGEs in *B. glabrata*

In attempts to decipher changes in the snail transcriptome in response to *S. mansoni* infection, several RNA profiling studies were undertaken. These included using Differential Display Reverse Transcriptase-Polymerase Chain Reaction (DD-RT-PCR) and generating Expressed Sequence Tags (ESTs) from snails before and after parasite infection. Interestingly, irrespective of which profiling tool was adopted, large numbers of transcripts encoding RT were identified.[10,21] These included transcripts showing sequence identity to the RT region of the sea squirt, *Ciona intestinalis* non-LTR- retrotransposon, *CiI* and to the silkmoth *Bombyx mori* retrotransposon, *mag*. Most importantly, from ESTs generated from the resistant snail hemocytes, the cellular component of the snail's innate defense system, a transcript (GenBank Acc. No. AW740305) corresponding to the original BGR2 MGE was identified.[10]

Since this fully sequenced cDNA transcript designated as *Bg37* (insert size 476 bp, GenBank Acc. No. EF413179, also used a probe in Fig. 1B) was found by qualitative RT-PCR to be expressed in both normal hemocytes and those exposed to miracidia, it was not initially considered to play a role in the defense reaction against parasite infection. However, by real-time RT-PCR differential regulation of this transcript upon parasite exposure was revealed (Fig. 2). As shown in Figure 2, early response (5 h post exposure) to the parasite demonstrated up-regulation of the *Bg37* transcript in both resistant (BS-90 and LAC) and susceptible (NMRI and M-line) snails. Furthermore, analysis over a 48 h time period post-exposure showed up-regulation of this RT transcript expression with susceptible snails (NMRI and M-line) showing higher levels of

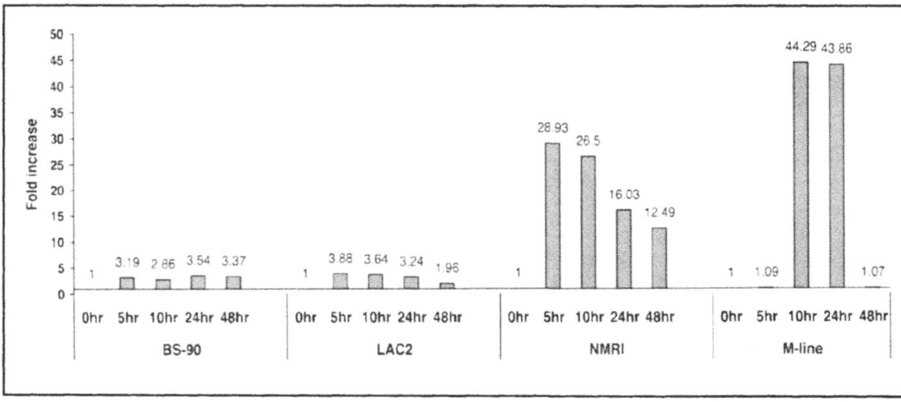

Figure 2. Real time PCR was done using an Applied Biosystems 7300 Real Time PCR System (Applied Biosystem, Foster City, CA). Reactions were performed in a one-step format with total RNA (80 ng). Synthesis of first strand cDNA and amplification by PCR were performed sequentially in a single tube using Full Velocity SYBR Green QRT-PCR Master mix according to the manufacturer's instructions (Stratagene). Reactions (25 μl final volume) contained the following; Bg37 Gene Specific Primers (200 nM) forward (5'-GCTCCATTAAACCGAACAGAC-3') and reverse (5'ATCACCGACCTACTTGCACC-3'); the housekeeping gene myoglobin primers (50 nM) forward (5'GATGTTCGCCAATGTTCCC-3') and reverse (5'-AGCGATCAAGTTTCCCCAG-3') used to adjust for concentration differences of reverse transcribed RNA; 300 nM of reference dye, 1x of Full Velocity SYBR Green QRT-PCR master mix containing RT-PCR buffer, SYBR green I dye, $MgCl_2$ and nucleotides. The amplification protocol included an initial incubation at 48°C for 45 min for cDNA synthesis and a 95°C initial denaturation for 10 min followed by 40 cycles with 95°C denaturation for 10 sec, and annealing/ amplification at 58°C for 1 min. Detection of the fluorescent product was carried out at the end of the amplification period. All amplifications were run in triplicate and the fluorescence threshold value (Ct) was determined using the 7300 System v1.3.1 SDS software (Applied Biosystems). Comparison of the expression of the RT encoding gene between pre and post-exposed snails was determined using delta-delta (ΔΔ) Ct. Results were transformed into 'fold increase' according to the following formula[37]:

Fold increase $= 2^{-\Delta\Delta Ct}$

$= 2^{-[(Ct_{RT,\ expose} - Ct_{myoglobin,\ expose}) - (Ct_{RT,\ un\text{-}expose} - Ct_{myoglobin,\ un\text{-}expose})]}$

Statistical analysis to determine standard error and 95% confidence intervals (CI) were calculated for each sample based on 'fold increase' results obtained using the above equation. In order to determine significant differences ($P < 0.05$) in expression of the RT encoding transcript at the different time points preand post-exposure for the snail stocks utilized (resistant, BS-90, LAC and susceptible, M-line, NMRI), the exposed CI value was subtracted from the mean unexposed 'fold increase' (1 = value for unexposed snails). 'Fold increase' > 1.0 was considered as significant.

induction after infection (from 28 to 44 fold) compared to the levels (up to 4 fold) observed in resistant snails (BS-90 and LAC) exposed for the same period of time. Results shown in Figure 2 are from three separate experiments done with size-matched snails (3 individuals at 4-6 mm in diameter) that were individually exposed to miracidia (5/snail).

Similar to the above results was our earlier finding that functional RT enzyme activity, measured in these snails before and after parasite exposure (0-72 h), displayed changes in activity levels depending on whether snails were resistant or susceptible to infection. Contrary to the aforementioned results, however, investigation of functional RT activity showed that whereas the enzyme level increased in parasite exposed resistant snails it remained steady and unchanged in susceptible snails exposed for the same period of time.[10] Taken together, these results (RT-PCR and functional enzyme analysis) provide evidence that regulation of certain classes of MGEs exhibits different paradigms with regard to transcription and translation in *B. glabrata* upon exposure to *S. mansoni*. The preliminary evidence that differential regulation of MGEs may be a factor in the snail host parasite relationship provided the impetus to isolate and characterize the complete full-length genomic copy of the active BGR2 related transcript, *Bg37*.

Identification of *nimbus* (*BgI*): BAC Clones Cross-Hybridizing to the Genomic BGR2 and cDNA Transcript *Bg37*

To further characterize the transcript *Bg37*, studies were undertaken to identify and sequence the complete full-length actively transcribed non-LTR retrotransposon. For this, we screened a BAC library constructed from the parasite resistant snail BS-90. Clones that hybridized to both ^{32}P-labeled genomic BGR2 and the transcript *Bg37* inserts were isolated. Positive clones were only processed further, if by nested PCR they revealed the expected size fragment within the BAC clones. From 10 positive BACs that cross-hybridized to both probes, five were subjected to preliminary sequencing and one clone (BRIBAC72bg_line5; GenBank Acc. No. EF418587) was fully sequenced (Fig. 3). The entire insert of this clone (65,764 bp) was found to contain one complete full-length element (element I; GenBank Acc. No. EF413180), two full-length elements with deletions and substitutions (elements II and III), and 10 truncated copies (elements A, B and C).[20] The intact non-LTR retrotransposable element corresponding to BGR2 and *Bg37* was named *nimbus*, after Harry Potter's magical flying broomstick,[22] for its potential of becoming mobile in the snail genome. In addition, the designation *BgI* has also been included in the naming of the snail element to signify in which clade it belongs (clade I). As illustrated in the graphical representation shown in Figure 3, aside from the *nimbus* (*BgI*) elements in this BAC clone, three other MGEs unrelated to *nimbus* (*BgI*) (one copy each of non-LTR retrotransposons related to *CRI* and *RTE* and two copies of a *gypsy*-like LTR retrotransposon, all truncated), were found to be present. The intact *nimbus* (*BgI*) elements contained the hallmark domains found in all Class I non-LTR retrotransposons; ORF 1 (encoding a nucleic acid binding protein) and ORF 2 encoding the apurinic/apyrimidinic endonuclease (APE), reverse transcriptase (RT) and RNase H (RNH).[17,23] Blastx[24] analysis of ORFs 1 and 2 showed significant matches (40-50%) to non-LTR retrotransposons from *Aedes aegypti* (*MosquI*),[25] *Drosophila melanogaster* (I-factor),[17,26] and *Ciona intestinalis* (CiI)[27] with E values of > e^{-20}.

Detailed nucleotide sequence analysis of the complete full-length *nimbus* (*BgI*) (5,869 bp; GenBank Acc. No. EF413180) has recently been reported[20] and summarized here as follows: ORF 1, also known as the Nucleic Acid Binding Domain (NABD) was found to contain three zinc finger domains of the CCHC type[28] normally seen in the same domain of retroviruses. These zinc finger domains were of the invariant ($CX_2CX_4HX_4C$) and variant ($CX_2CX_4HX_6C$) types. Upstream of the ORF 1 initiation codon, putative promoter sequences TATA (-112) and CAAT (-165) were detected. Twenty nucleotides separated ORF 1 from the start codon of ORF 2. Noncoding Regions (NCR) flanked the 5' and 3' ends of all the *nimbus* (*BgI*) copies (full-length and truncated) present within the BAC clone. Features found in the flanking 5' NCR (565 bp) and 3' NCR (234 bp), although complex, also

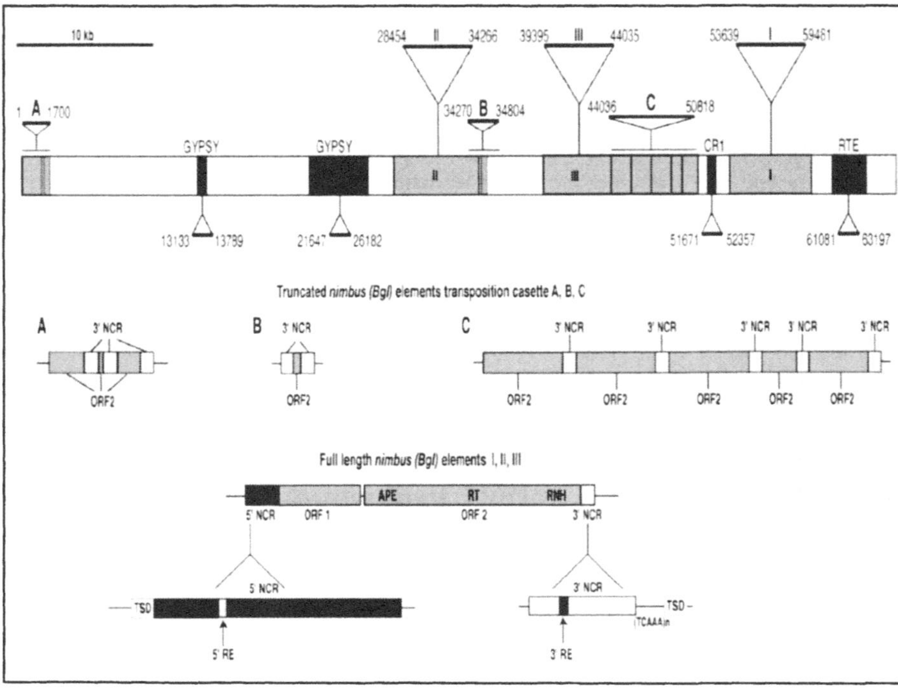

Figure 3. A schematic representation of the *B. glabrata* BAC BRIBAC72bg-line5 (GenBank Acc. No. EF418587) containing multiple copies of the transposable element *nimbus* (*Bgl*). The 65,764 bp BAC insert contained one full-length-complete *nimbus* (*Bgl*) element (Element I), 2 full-length but incomplete *nimbus* (*Bgl*) elements (Elements II and III) and 10 truncated copies that are present in multiple transposition cassettes (Elements A, B, and C). All the *nimbus* sequences are shown in dark grey. Truncated copies of other non-LTR retrotransposons CR1 and RTE (black) and the LTR retrotransposon *gypsy* (black) were also present in the BAC sequence. A 10 kb bar scale line indicates the relative length of the BAC clone. The beginning and end position of each of the transposable elements within the BAC clone is indicated. The detailed organization of the full-length elements (I, II and III) is shown at the bottom of this Figure. Each element contains 2 ORFs (ORF1 and ORF2) separated by a small intervening sequence. ORF2 contains 3 domains (APE, RT and RNH). Preceding ORF1 is the 5' NCR whose expanded schematic is shown in black directly below it. This region consists of a terminal 5'TSD (white), the conserved sequences of the 5' NCR (black) with a 16 bp 5' RE (white). Following ORF2 is the 3'NCR whose expanded schematic is shown in white directly below it. This region consists of a terminal 3' TSD (white), the conserved sequence of the 3' NCR (white) with its 16 bp 3' RE (shown in black) and separated from the 3'TSD by the variable tandem $TCAA_n$ or CAA_n repeats. The truncated copies of *nimbus* are present as transposition cassettes (A: 3 copies; B: 2 copies and C: 5 copies). The detailed organization of these truncated transposition cassettes are shown in panels A, B and C (grey boxes). Reprinted from: Raghavan N et al. Int J Parasitol 2007; doi:10.1016/j.ipara. 2007.04.002;[20] with permission from Elsevier, ©2007.

provided the most novel information about this snail element compared to other known MGEs of the same class. These included conserved Repeat Elements (RE), 16 bp in length that showed two nucleotide substitutions from each another. The function of this unique feature in the snail element remains unknown. Moreover, unlike other elements, the flanking ends of ***nimbus*** (*Bgl*) do not end in either poly (A) or 'TAA' tandem repeats, but rather the 3' NCR ends with the variable repeat sequence $TCAA_n$ or CAA_n. These repeats are AT rich (75%) and define the junction of the 3' Target Duplication Site (TSD) at the end of each element. Putative TSDs[29] in the full-length *nimbus* (*BgI*) elements were found to be variable (in size and content) and may be generated during the asymmetric cleavage of the target DNA sequence by the *nimbus* (*BgI*) endonuclease (APE). Phylogenetic analysis using the *nimbus* (*BgI*) RT sequence revealed that the snail element belongs to the I-clade of non-LTR retrotransposons and is closely related to *CiI* of *Ciona intestinalis*,[27] *mosquI* of *Aedes aegypti*[25] and *both L1Tc* and *INGI* of trypanosomes.[30-31]

Detection of *nimbus* (*BgI*) Related Sequences in Schistosomes

To determine whether sequences corresponding to the snail *nimbus* (*BgI*) element may be present in the parasite genome, we analyzed the schistosome databases using various regions of the *nimbus* (*BgI*) sequence. Intriguingly, in addition to the coding regions of *nimbus* (*BgI*), the *nimbus* 5' and 3' NCRs have been discovered to be highly conserved in the genome of *S. mansoni* when screened against the *S. mansoni* whole genome shotgun database available either at the Sanger Center using BLASTN or NCBI using discontiguous Mega BLAST. Significant nucleotide matches (>85% identity, Table 1) between the *nimbus* (*BgI*) 5' and 3' NCRs and parasite might be an indication of possible horizontal transfer of this sequence between the snail host and the parasite. While we do not have a one to one match of the flanking sequences with *S. japonicum*, some of the *S. mansoni* shotgun sequences that show sequence similarity to our flanking sequences also have matches to *S. japonicum*. This may simply indicate that the shotgun sequence of the *S. mansoni* clone, while showing sequence identity to our flanking region, may also exhibit sequence identity to *S. japonicum* sequence at a different region within the same clone and in fact may represent a transposition hot spot. Primers based on the snail *nimbus* (*BgI*) 5' and 3' NCRs have been designed and used in PCR and RT-PCR amplifications of *S. mansoni* genomic DNA and mRNA, respectively. Both reactions gave products of the expected size (data not shown)

Table 1. Discontiguous Mega BLAST analysis of the representative* nimbus (BgI) *5' and 3' NCRs against the NCBI* Schistosoma mansoni *WGS database

Homology of *nimbus* (*BgI*) 5' NCR to *S. mansoni* WGS sequences:			
GenBank Acc. No.	**Description**	**% Identity**	**E-Value**
284173464	shisto4591f09.p1k	100%	8e-86
267992943	shisto174e09.q1k	87%	7e-07
267884589	shisto217e09.q1k	87%	7e-07
267991264	shisto217e09.q1kw	100%	3e-06
Homology of *nimbus* (*BgI*) 3' NCR to *S. mansoni* WGS sequences:			
GenBank Acc. No.	**Description**	**% Identity**	**E-Value**
275050373	shisto3280a11.q1k	93%	4e-86
284171603	shisto4444b01.p1k	90%	2e-27

Significant matches (>85% identity) were obtained with both the 5' and 3' NCRs against the *S. mansoni* whole genome shotgun sequences. The data shown in the table denotes the GenBank accession numbers of the *S. mansoni* sequences, their descriptive name, % identity and corresponding E-values to the *nimbus* (*BgI*) 5' and 3'NCRs.

indicating the presence and expression of conserved *nimbus* (*BgI*)-like MGE sequences between the snail and the parasite. Although several MGEs have been characterized from the schistosomes (*S. mansoni* and *S. japonicum*),[32] none so far have been identified as belonging to the I-clade. The presence of highly conserved flanking sequences between the snail and the parasite and their possible role in the host/parasite relationship, including horizontal transfer during the evolutionary process, requires further investigation.

Chromosomal Location of *nimbus* (*BgI*) in *B. glabrata* by Fluorescent In Situ Hybridization (FISH)

In order to visualize the distribution of the *nimbus* (*BgI*) element in the snail genome, we performed Fluorescence In Situ Hybridization (FISH). To standardize and optimize conditions for this procedure, the immortalized embryonic snail cell-line Bge,[33] derived from *B. glabrata* snails, was used as the source of material to generate high quality metaphase chromosomes for hybridizations. The BAC containing *nimbus* (*BgI*) was employed as the probe. This Bge cell line is convenient for the preparation of plentiful and reproducible metaphase spreads. But, as with many cell lines its karyotype is abnormal and most likely the result of having been in culture for many years (Fig. 4). Thus, instead of the normal *ex vivo* karyotype of 36 chromosomes,[34] there are 66 chromosomes, on average, per metaphase spread generated from Bge cells. What event(s) produced the abnormal chromosome amplification/duplication seen in these cells remains unknown. Figure 4A displays an image of a DAPI-banded preparation of chromosomes that has been made grayscale and inverted. It is possible to see that the chromosomes range in size (2 - 6 μm) with many micro-chromosomes present, and there are banded areas of lighter and darker staining that may help to karyotype the chromosomes of this cell line.

In order to determine the genomic distribution of the *nimbus* (*BgI*) element, the BAC containing the *nimbus* (*BgI*) was subjected to a nick translation reaction incorporating biotinylated nucleotides. The labeled probe was codenatured with the metaphase chromosomes on glass slides. There were no repetitive suppressing sequences used in the process since (1) a method for FISH not requiring suppression has been developed[35] and (2) since *nimbus* (*BgI*) is a repetitive element, the signal may be lost due to competitive suppression. Figure 4B demonstrates a typical distribution of the BAC containing *nimbus* (*BgI*) in red. The *nimbus* (*BgI*) probe signal was not observed on all chromosomes (see smaller arrows) but in the majority of chromosomes where a FISH signal was evident it was mainly distributed at the centromeric and subcentromeric regions of the chromosomes (see larger arrows). This region of the chromosomes is often associated with repetitive DNA and is a rapidly evolving region in genomes.[36] As we discover more about the organization of the *B. glabrata* chromosomes, both in the cell line and ex vivo, it will become possible to map more carefully the exact bands and chromosomal regions where *nimbus* (*BgI*) resides.

Conclusions and Future Directions

Modern genome biology may help in developing new molecular tools to combat schistosomiasis. For the three organisms that are pertinent to transmission of this disease—the parasite, the intermediate snail host, and the definitive human host—comparative analysis of their genomes, especially the snail-host and parasite, could potentially lead to experiments that manipulate gene expression sufficiently to translate into practical tools for disrupting parasite transmission. With the *nimbus* (*BgI*) sequence we can now suggest that both at the bioinformatics and experimental levels, there may be sequence sharing between the intermediate host and the parasite. This may be an important glimpse into the complex regulation of gene expression that underlies the outcome (resistance/susceptibility) of the snail/schistosome encounter. Thus, *nimbus* (*BgI*), a mobile element of the freshwater snail *B. glabrata* and the first full-length non-LTR MGE described from a mollusk, may provide further insights into the snail/parasite relationship and, be useful for genetic transformation studies aimed towards reducing snail-borne diseases.

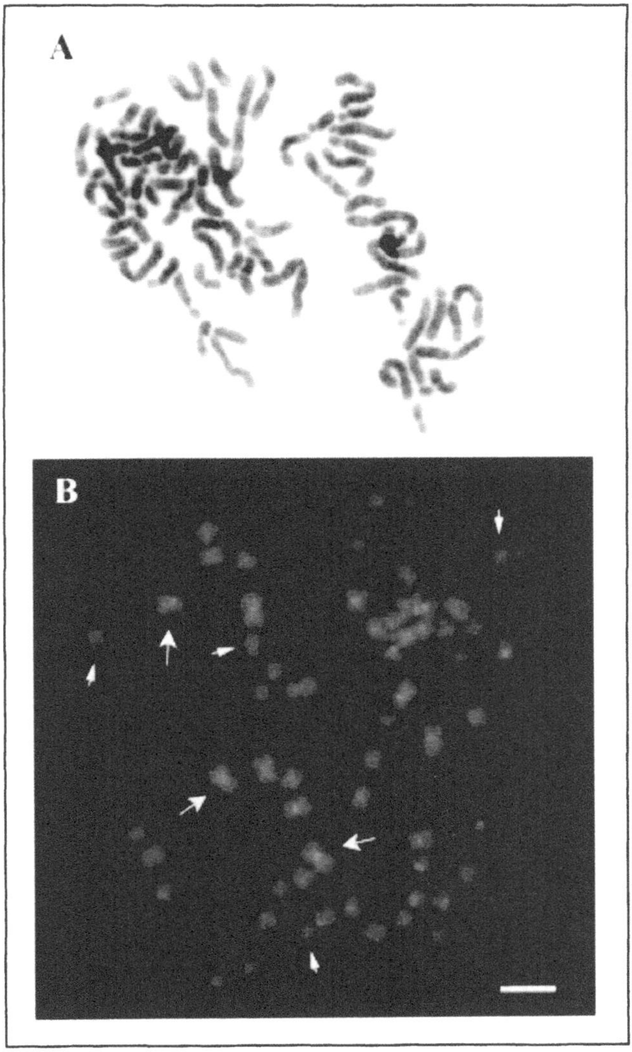

Figure 4. Chromosome preparations from the immortalized *B. glabrata* embryonic cell line Bge was prepared by incubating the cells in colcemid for 2 h to arrest their growth at metaphase. Cells were harvested and subjected to hypotonic treatment before fixing in methanol: acetic acid (1:1). In panel A, a typical chromosome spread is displayed where the chromosomes have been stained with the DNA intercalating dye 4', 6-diamidino-2-phenylindole (DAPI). The grayscale image has been inverted to allow better visualization of the DAPI light and dense banding patterns. Panel B displays a Bge chromosome spread that has been subjected to FISH with the BAC DNA containing the *nimbus* (*BgI*) sequence as a probe. The BAC DNA has been labeled with biotinylated nucleotides during a nick translation reaction. After denaturation of the probe and sample, the hybridized probe is delineated using Cyanine-3 conjugated to streptavidin. The blue staining is the pseudo-colored DAPI image. The images were taken using a Photometrics CCD camera on a Zeiss Axiophot microscope with Smartcapture 2 software for imaging. The large arrows indicate the chromosomes that display major areas of *nimbus* (*BgI*) staining seen at or near the centromeric regions. The smaller arrows indicate chromosomes that have none, or very little signal. The scale bar = 5 μm. A negative control was used whereby a BAC containing a chicken sequence was used in a parallel hybridization. No signal was observed on the snail chromosomes using the nonspecific probe (not shown).

Acknowledgements

We thank Dr. Hervé Tettelin, Jessica Hostetler and Luke Tallon for their helpful collaboration in bringing the full sequence of the *nimbus* (*BgI*) BAC to fruition. We also thank Dr. Peter FitzGerald for his encouragement and help with the bioinformatics undertaken for this project, and Dr. Fred Lewis for his support and help with editing this manuscript. We wish to acknowledge the Biowulf PC/Linux cluster at the NIH, Bethesda, MD (http://biowulf.nih.gov) used for the sequence analysis. This work was supported by awards from the USF-Sandler Foundation (9000005939) and NIH-NIAID (AI-063480-01).

References

1. Fedoroff NV. Transposable elements as a molecular evolutionary force. Ann NY Acad Sci 1999; 870:251-264, (Review).
2. Coates CJ, Jasinskiene N, Miyashiro L et al. Mariner transposition and transformation of the yellow fever mosquito, Aedes aegypti. Proc Natl Acad Sci USA 1998; 95:3748-3751.
3. Jasinskiene N, Coates CJ, Benedict MQ et al. Stable transformation of the yellow fever mosquito, Aedes aegypti, with the Hermes element from the housefly. Proc Natl Acad Sci USA 1998; 95:3743-3747.
4. Boulo V, Cadoret JP, Shike H et al. Infection of cultured embryo cells of the pacific oyster, Crassostrea gigas, by pantropic retroviral vectors. In Vitro Cell Dev Biol Anim 2000; 36:395-399.
5. Jordan P, Christie JD, Unrau GO. Schistosomiasis transmission with particular reference to possible ecological and biological methods of control. Acta Trop 1980; 37:95-135.
6. Fenwick A, Webster JP. Schistosomiasis: Challenges for control, treatment and drug resistance. Curr Opin Infect Dis 2006; 19:577-582.
7. Richards CS. Genetics of a molluscan vector of Schistosomiasis. Nature 1970; 227:806-810.
8. Lewis FA, Patterson CN, Knight M et al. The relationship between Schistosoma mansoni and Biomphalaria glabrata: Genetic and molecular approaches. Parasitology 2001; 123(Suppl):S169-179.
9. Schneider O, Zelck UE. Differential display analysis of hemocytes from schistosome-resistant and schistosome-susceptible intermediate hosts. Parasitol Res 2001; 87:489-491.
10. Raghavan N, Miller AN, Gardner M et al. Comparative gene analysis of Biomphalaria glabrata hemocytes pre and post-exposure to miracidia of Schistosoma mansoni. Mol Biochem Parasitol 2003; 126:181-191.
11. Nowak TS, Woodards AC, Jung Y et al. Identification of transcripts generated during the response of resistant Biomphalaria glabrata to Schistosoma mansoni infection using suppression subtractive hybridization. J Parasitol 2004; 90:1034-1040.
12. Lockyer AE, Spinks J, Noble LR et al. Identification of genes involved in interactions between Biomphalaria glabrata and Schistosoma mansoni by suppression subtractive hybridization. Mol Biochem Parasitol 2007; 151:18-27, (Epub 2006 Oct 20).
13. Bouchut A, Coustau C, Gourbal B et al. Compatibility in the Biomphalaria glabrata/Echinostoma caproni model: New candidate genes evidenced by a suppressive subtractive hybridization approach. Parasitology 2007; 134:575-588 (Epub 2006).
14. Raghavan N, Knight M. The snail (Biomphalaria glabrata) genome project. Trends Parasitol 2006; 22:148-151.
15. Knight M, Miller A, Raghavan N et al. Identification of a repetitive element in the snail Biomphalaria glabrata: Relationship to the reverse transcriptase-encoding sequence in LINE-1 transposons. Gene 1992; 118:181-187.
16. Fanning TG, Singer MF. LINE-1: A mammalian transposable element. Biochim Biophys Acta 1987; 910:203-212, (Review).
17. Fawcett DH, Lister CK, Kellett E et al. Transposable elements controlling I-R hybrid dysgenesis in D. melanogaster are similar to mammalian LINEs. Cell 1986; 47:1007-1015.
18. Eickbush TH. Origin and evolutionary relationship of retroelements. In: Morse SS, ed. Evolutionary Biology of Viruses. Raven Press, 1994:121-157.
19. Malik HS, Burke WD, Eickbush TH. The age and evolution of non-LTR retrotransposable elements. Mol Biol Evol 1999; 16:793-805.
20. Raghavan N, Tettelin H, Miller A et al. Nimbus (BgI): An active non-LTR retrotransposon of the Schistosoma mansoni snail host Biomphalaria glabrata. Int J Parasitol 2007; doi:10.1016/j.ipara.2007.04.002.
21. Knight M, Ongele O, Lewis FA. Molecular studies of Biomphalaria glabrata, an intermediate host definitive of Schistosoma mansoni. Int J Parasitol 2000; 30:535-541.
22. Rowling JK. Harry Potter and the Sorcerer's Stone. Scholastic Press, 1998.

23. Feng Q, Schumann G, Boeke JD. Retrotransposon R1Bm endonuclease cleaves the target sequence. Proc Natl Acad Sci USA 1998; 95:2083-2088.
24. Altschul SF, Gish W, Miller W et al. Basic local alignment search tool. J Mol Biol 1990; 215:403-410.
25. Tu Z, Hill JJ. MosquI, a novel family of mosquito retrotransposons distantly related to the Drosophila I factors, may consist of elements of more than one origin. Mol Biol Evol 1999; 16:1675-1686.
26. Berezikov E, Bucheton A, Busseau I. A search for reverse transcriptase-coding sequences reveals new non-LTR retrotransposons in the genome of Drosophila melanogaster. Genome Biol 2000; 1:RESEARCH0012, (Epub 2000).
27. Permanyer J, Gonzàlez-Duarte R, Albalat R. The non-LTR retrotransposons in Ciona intestinalis: New insights into the evolution of chordate genomes. Genome Biol 2003; 4:R73.
28. Apweiler R, Attwood TK, Bairoch A et al. The InterPro database, an integrated documentation resource for protein families, domains and functional sites. Nucleic Acids Res 2001; 29:37-40.
29. Finnegan DJ. Transposable elements: How non-LTR retrotransposons do it. Curr Biol 1997; 7:R245-248, (Review).
30. Kimmel BE, ole-MoiYoi OK, Young JR. Ingi, a 5.2-kb dispersed sequence element from Trypanosoma brucei that carries half of a smaller mobile element at either end and has homology with mammalian LINEs. Mol Cell Biol 1987; 7:1465-1475.
31. Martin F, Maranon C, Olivares M et al. Characterization of a nonlong terminal repeat retrotransposon cDNA (L1Tc) from Trypanosoma cruzi: Homology of the first ORF with the ape family of DNA repair enzymes. J Mol Biol 1995; 247:49-59.
32. Brindley PJ, Laha T, McManus DP et al. Mobile genetic elements colonizing the genomes of metazoan parasites. Trends Parasitol 2003; 19:79-87, (Review).
33. Hansen EL. A cell line from embryos of Biomphalaria glabrata (Pumonata): Establishment and characteristics. In: Maramorosch K, ed. Invertebrate Tissue Culture. New York: Academic Press, 1976:785-799.
34. Goldman MA, LoVerde PT, Chrisman CL et al. Chromosomal evolution in planorbid snails of the genera Bulinus and Biomphalaria. Malacologia (Int J Malacol) 1984; 25:427-446.
35. Masabanda JS, Griffin, DK. Generation of chromosome paints: Approach for increasing specificity and intensity of signals. Biotechniques 2003; 34:530-536.
36. Henikoff S, Ahmad K, Malik HS. The centromere paradox: Stable inheritance with rapidly evolving DNA. Science 2001; 293:1098-1102, (Review).
37. Livak KJ, Schmittgen TD. Analysis of relative gene expression data using real-time quantitative PCR and the 2 (-Delta Delta C [T]) method. Methods 2001; 25:402-408.

CHAPTER 5

Transposon Mediated Transgenesis of Mosquitoes

Craig J. Coates*

Abstract

The first six years of mosquito transgenesis have seen dramatic increases in the number of mosquito species that have been genetically transformed, the number of transposable element vectors that have been utilized to produce transformants, and in the diversity of applications for which the basic technology has been used. While there have been many encouraging advances, there have been some problems encountered and the frequency by which independent transgenic lines of mosquitoes can be generated, remains relatively low. Even with these issues remaining to be resolved, the ability to genetically transform mosquito species has seen the utilization of a wide range of genetic constructs for the purposes of gene discovery, unlocking the keys to gene regulation and the production of transgenic strains that are refractory to certain pathogens. This chapter will address the advances, issues and provide a view to the future of mosquito transgenesis.

Introduction

Transposon mediated transgenesis of insect species originated over twenty years ago with the use of the *P* transposable element to genetically transform the vinegar fly, *Drosophila melanogaster*.[1] The ease and efficiency by which large numbers of independent transgenic lines could be generated further expanded the utility of what was an already well-established model genetic organism and likely played a significant role in the selection of this species for a whole genome sequencing effort.[2] With a minimal amount of training and specialized equipment, the production of transgenic lines of *D. melanogaster* can be a routine experimental procedure in essentially any laboratory, producing appreciable numbers of transformed lines from each experiment.

Unfortunately, the ease and efficiency of the *P* element system for *Drosophila* transformation may have ultimately hampered efforts to transform nondrosophilid insects. As was the most logical progression, many researchers attempted, unsuccessfully, to utilize the *P* element transformation system in toto, for transforming their species of choice. However, it was ultimately demonstrated that the activity of the *P* element is limited to a restricted group of *Drosophila* species, having been found to be essentially inactive in many nondrosophilid species.[3]

Renewed efforts in the early to mid 1990s to identify alternative insect transposable elements for use as genetic transformation vectors yielded a number of candidates in a diverse range of species. The use of plasmid-based transposition assays[4] to confirm the activity of a given transposon in a particular insect species rapidly provided active transposons for testing in germline transformation experiments.

*Craig J. Coates—Department of Entomology, Texas A&M University, 2475 TAMU, College Station, Texas 77843-2475, USA. Email: c-coates@tamu.edu

Mobile Genetic Elements in Metazoan Parasites, edited by Paul J. Brindley.

In addition to the activity of the *P* element, experimenters in the *Drosophila* system were also blessed with an abundance of visible mutant strains and functional gene constructs to enable the rescue of these mutations. This provided the essential marker genes needed for the identification of transgenic individuals when these experiments were performed in the mutant background. Similarly, the first two examples of nondrosophilid transformation also utilized visible marker genes to rescue a mutant eye color phenotype,[5,6] a luxury not available for many species. The advent of the green fluorescent protein (GFP),[7] a nondestructive visible marker gene with no host cell requirements, provided the impetus for the dramatic increase in the number of mosquito species transformed, as there were now available a multitude of transposon vectors and a universal marker gene.

While the number of mosquito species that have been transformed has increased, the frequency by which the transgenic lines are produced has not increased dramatically and is particularly low when considering the difficulty, labor and cost involved in rearing an insect species that has different environmental conditions for each life stage and a blood meal requirement for egg production. A number of aberrant and unstable integration events have been observed in some transgenic mosquito lines. Furthermore, it is clear that the expression profiles of the integrated transgenes are subject to the influence of the surrounding chromatin environment, leading to variability between independent lines. A number of research avenues are being pursued to mitigate these issues.

The future of mosquito transgenesis will lead to the utilization of additional transposons, the application of endogenous regulatory sequences to enable germ-line expression of transposases, the manipulation of host genome defense mechanisms and the incorporation of endogenous boundary elements to improve transgene expression. In the short-term, the number of transgenic mosquitoes that are being produced continues to increase, as do the multitude of applications for which this technology is being exploited, including investigations of gene function and regulation, and for the purposes of reducing pathogen transmission.

Binary System for Mosquito Transgenesis

The system used for mosquito transgenesis is binary in nature in that two plasmids are used. One plasmid contains the transposase-coding region, which is typically regulated by a strong constitutive promoter, thus enabling high expression levels when micro-injected into preblastoderm embryos. This construct typically contains minimal additional transposon sequences, in particular lacking one or both inverted terminal repeats so that this construct cannot be mobilized to the chromosomes. It is thus referred to as the helper plasmid, providing the transposase for mobilizing in trans, an additional genetic construct.

The second plasmid is called the donor plasmid as it contains the genetic elements that will be integrated into the host chromosomes. It consists of a usually internally deleted transposon such that it contains the inverted terminal repeats and other internal sequences required for transposition, yet does not itself have the potential to encode for the transposase, thus relying on the helper plasmid to provide the transposase for mobilization. This ensures that when the donor element is integrated into the host chromosomes it will remain stably inserted, as it cannot remobilize itself. It is important to determine that the host genome does not contain related transposons that might be capable of cross-mobilizing the donor element, causing unstable integrations. The donor element also contains the marker gene that is used for the identification of the transgenic individuals, the expression of which is either controlled by a constitutive promoter or in many cases in mosquito transformation, a synthetic promoter based on a consensus sequence for an evolutionarily conserved regulatory element involved in eye development;[8] (Fig. 1). This has been particularly successful in all species tried to date. Finally, the transgene and associated regulatory elements are also included in the donor element.

There are other alternatives available for the provision of helper transposase, including the use of purified recombinant protein,[9] or in vitro transcribed RNA.[10] The logic behind these alternatives is to remove concerns over promoter utilization and also to enable integration

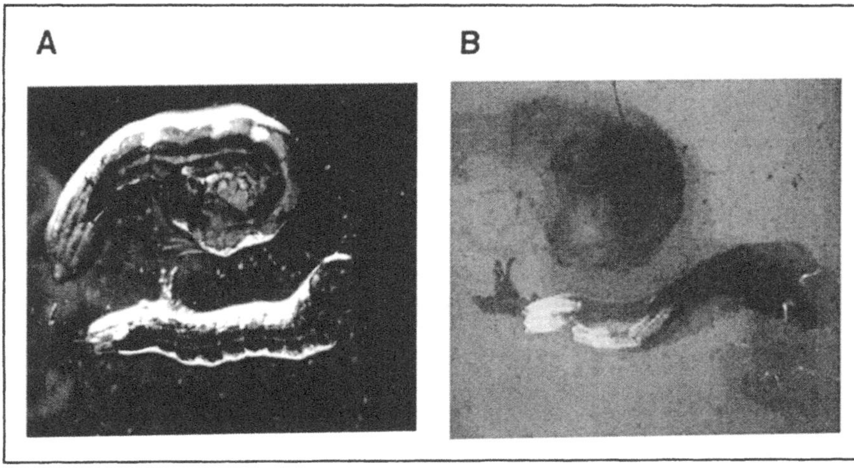

Figure 1. EGFP fluorescence in transgenic *Aedes aegypti* mosquitoes. A) Bright field image of a transgenic pupa and larva. B) Fluorescent image showing EGFP expression. Note that even though an eye-specific promoter was used to control the expression of the EGFP marker gene, in addition to the two fluorescent eyes seen in the larva, there is also additional fluorescence detected in the body, presumably as a result of a position effect caused by regulatory sequences flanking the transgene.

events to occur earlier in the development of the germline, in the hope of producing transgenic lines in which higher proportions of the progeny are transgenic. The number and nature of the integrations in the transgenic individuals are typically confirmed through the use of Southern Blot (Fig. 2) and inverse polymerase chain reaction (IPCR) analysis.

Table 1, provides a summary of the mosquito species that have been genetically transformed and the transposons used as gene vectors. The most obvious feature of the table is the prevalence of transformations performed in *Ae. aegypti* with different transposons. This imbalance would be even more substantial if all genetic transformation experiments were listed, rather than just the first instance in each species with each element. In addition to its obvious importance as a vector of both Dengue and Yellow Fever viruses, the yellow fever mosquito, *Ae. aegypti*, is also relatively easy to rear and manipulate in a laboratory setting, which provides a significant advantage for transgenesis of this species relative to many other mosquito species. Similarly, more genetic transformation experiments have been reported to date in *An. stephensi* (~10 genetic constructs) compared to *An. gambiae* (2 constructs), due in great part to the comparative ease of rearing, crossing, egg manipulation and survival following microinjection. The basic and in particular, reproductive biology of each mosquito species will continue to have a dramatic effect on the number of genetic transformation experiments and thus transgenic lines that are produced, particularly for those species that are difficult to culture and have complex mating system requirements. While not listed on the table, it is also worth noting that only the first two genetic transformation experiments referenced utilized an eye color gene marker rescue of a mutant line, the remainder all relied on the use of a fluorescent marker gene.

The majority of mosquito transgenesis experiments have reported transformation rates of less than 10% and in many cases, closer to 1% of G_O adults produce transgenic progeny. In addition, the proportions of the progeny from these individuals that are transgenic are also typically very low. Based on these current frequencies, it appears likely that there will need to be a quantum leap in the frequency and efficiency of the production of transgenic lines for this technology to be of routine use in a large number of laboratories. Such an advance may take the form of an alternative transposable element system with higher frequencies of integration, or

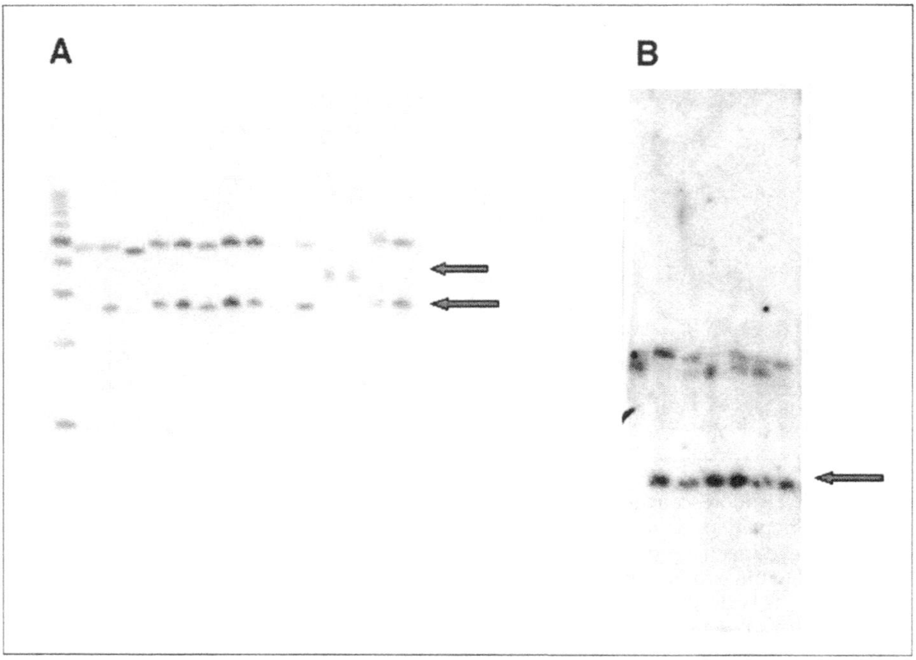

Figure 2. Southern blot analysis of transgenic mosquitoes. A) *Eco*RI digested genomic DNA from independent transgenic lines and siblings, hybridized with a radio-labeled EGFP-encoding fragment. The arrows indicated the presence of unique hybridizing fragments, which are indicative of independent insertion events. B) *Eco*RI digested genomic DNA from independent transgenic lines hybridized with a radio-labeled cecropin transgene fragment. The arrow indicates the presence of a constant internal cecropin gene fragment.

the modification of existing systems to maximize the number of germline integration events that are occurring, as well as the recovery of an increased number of functional integration events that result in suitable and stable levels of transgene expression.

Applications of Genetic Transformation Technology in Mosquitoes

It is perhaps unlikely that mosquito gene discovery efforts based on insertional mutagenesis and enhancer trapping will occur on the scale utilized for *D. melanogaster*,[18] simply due to the cost, space and labor required to rear and maintain large numbers of mosquito lines. For example, over 50 *Drosophila* lines can be reared in the space required to rear a single mosquito line. Generational propagation of *Drosophila* only requires the transfer of new adults to a fresh bottle, whereas for mosquitoes this requires a blood meal, egg collection and maturation, hatching and larval rearing, pupal collection and water removal, often in a different container for each life stage. Regardless, efforts towards the production of the tools required for gene discovery have been proceeding in mosquito species, including the production of 'jumpstarter lines' in which the transposase source is integrated onto a mosquito chromosome such that a genetic cross can be used to remobilize donor elements to new genomic locations rather than repeated microinjection experiments.[19,20] Unfortunately, while functional transposase-producing lines have been made, there has to date been a lack of efficient germ-line remobilization detected, which has hampered progress in this area. Success in overcoming the current limitations would allow the use of insertional mutagenesis and enhancer trapping experiments and while these

Table 1. Mosquito species transformed and transposons utilized

Mosquito Species	Transposon[a]	Reference[b]
Aedes aegypti	*Hermes* (*Musca domestica*)	6
	MosI (*Drosophila mauritiana*)	11
	piggyBac (*Trichoplusia ni*)	8
	Tn5 (*Escherichia coli*)	12
Culex quinquefasciatus	*Hermes*	13
Anopheles stephensi	*Minos* (*Drosophila hydei*)	14
	piggyBac	15
Anopheles gambiae	*piggyBac*	16
Anopheles albimanus	*piggyBac*	17

[a] The species of origin for the transposons is provided in parentheses the first time listed.
[b] Only the first instance of genetic transformation for each transposon in each species is referenced.

may not proceed on the scale used for *D. melanogaster*, they would allow for small scale focused studies of a particular tissue, developmental stage, or process.

A number of tools for investigations of gene function have been adapted for use in mosquito transgenesis. These include the phage P1 *creloxP* system, which can be used for manipulating integrated gene constructs to invert or delete delineated regions of genes and/or regulatory regions.[21] The Tet-On and Tet-Off systems also provide potent means by which to conditionally regulate gene expression in transgenic mosquitoes.[22] Finally, the use of RNA interference (RNAi) generating constructs in transgenic mosquitoes adds another powerful tool for the genetic manipulation of these species.[23]

Significant progress has been made in several investigations of mosquito gene regulation. In line with the prominent role and importance of the midgut, salivary glands and thoracic musculature in the infection and transmission of a variety of pathogens, most transgenic studies of mosquito gene regulation have been focused on these tissues.[24-27] Mosquito transgenesis has also been used to great effect to study the molecular basis of vitellogenin gene expression.[28]

The driving force for mosquito transgenesis research is the impact that these species have as vectors of pathogens of human and veterinary health importance. A number of experiments have been performed in the quest for the production of transgenic lines that are impaired in their abilities to transmit particular pathogens. A variety of strategies have been employed, including the use of enzymes, RNAi constructs, single chain antibodies and small peptides, all designed to interfere with some aspect of pathogen infection, spread, development or transmission.[27,29-32]

The production of these transgenic lines provides the proof of principle that mosquito transgenesis can be used to produce strains of genetically modified mosquitoes that are no longer able to transmit pathogens of human and agricultural importance. The next step in this process is the demonstration of an ability to spread the refractory transgenes through wild populations. One of the proposed strategies to achieve this goal is to use the inherent ability of transposons to increase in copy number in a genome and also to spread throughout a population. However, while the total number of experiments attempted to date is small, the results have not been overly promising in that a number of transposons have not been able to be remobilized in the mosquito germline, even though in some instances they have been efficiently remobilized in other insect genomes.[19,20,33-35]

One hypothesis is that the problem may be a lack of promoter activity in the mosquito germline, which has triggered efforts to clone genes that display germline-specific activity profiles, such that their regulatory sequences can be coopted for the purposes of transposase expression (personal communication, Anthony James, UCI; Coates, unpublished). It is also possible that the mosquito germline possesses defense mechanisms, such as RNAi or chromatin methylation, that are limiting the production of transposase transcripts in these cells. The continued search and testing of additional transposable elements may reveal a transposon that is less subject to down-regulation in the germline, perhaps being more likely for those that are endogenous to mosquito genomes.

The relative fitness of the transgenic individuals is also a primary factor of importance to these studies and conflicting reports have been presented with regards to this issue.[36-38] Further issues such as the genetic compatibility of 'laboratory' strains and wild-type populations remain to be addressed. It is also worth nothing that there are a number of other gene-drive mechanisms being investigated that are beyond the scope of this chapter.[39-41]

Issues in Mosquito Transformation

While the majority of reported transgenic experiments have resulted in the desired phenotype, there have been a number of examples whereby the integration events were not of the expected form, in that not all of the transposon construct was integrated, some of the donor plasmid backbone was also integrated, multiple tandem arrays of the plasmid or transposon were integrated and also combinations of the above.[9,20,42] While for the most part these aberrations have not dramatically limited the utility of the transgenic lines produced, these results do have some implications for the use of transposons to spread refractory genes.

Not surprisingly, position effects have also been shown to have an impact on transgene expression, such that independent lines of the same integrated construct display vastly different levels and even patterns of expression, presumably as a result of the influence of the surrounding regulatory elements of the endogenous genes. These can also be displayed in a sex- or tissue-specific manner. There have also been observations of position-effect variegation, whereby transgenic siblings display different expression levels. This is presumably as a result of the integration of the construct near the boundary of euchromatin and heterochromatin, resulting in the inactivation of the transgene in cells in which that genomic region has become embedded in heterochromatin, as these boundaries are not precise. This is of particular issue in a species such as *Ae. aegypti* in which a large proportion of the genome is heterochromatic and furthermore, that this is interspersed throughout the genome rather than just being concentrated at the telomeres and centromeres.

It is not clear which primary factors are behind the relatively low genetic transformation efficiencies observed for mosquito species. Certainly the manipulation of eggs is an issue in terms of allowing for sufficient quantities of DNA to be injected into all eggs that survive the procedure, as well as the relative lack of information concerning mosquito embryogenesis and germline development. There is also the potential that germline-specific expression is limited by the use of currently available promoters for driving transpose expression and the possibility that the mosquito genome possesses means to down-regulate transposase activity in these critical cells.

Future of Mosquito Transformation

In the near future it is likely that more species of mosquitoes will be genetically transformed. Many of these experiments may utilize additional transposons, particularly those that have been identified as active elements in mosquito genomes. Once identified, it seems reasonable to expect that any down-regulatory effects of the mosquito genome can be ameliorated to some degree, such that integration can occur at higher frequencies. The use of endogenous germline-specific regulatory sequences will potentially result in increased transgensis efficiencies and may lead to germline remobilization. The incorporation of endogenous boundary elements will assist in providing more consistent transgene expression, enable more efficient analysis of regulatory sequences and may result in an increase in the recovery of 'useful' transgenic lines. It is clear that efforts to understand the basic biology of transposons and the mechanisms that control and regulate their expression, movement and targeting will continue to be critical to efforts to improve the efficiency of mosquito transgenesis. That being said, even with the currently available sub-optimal systems, many candidate genes and their regulatory elements will be fully characterized, a plethora of refractory strains utilizing an increasing diversity of pathogen transmission blocking mechanisms will be developed and it is likely that remobilization of these genetic constructs will be achieved. While beyond the scope of this chapter, it is also hoped that it will eventually be possible to specifically target integration events to desired genomic regions, or even to individual genes.

References

1. Rubin GM, Spradling AC. Genetic transformation of Drosophila with transposable element vectors. Science 1982; 218:348-53.
2. Adams MD, Celniker SE, Holt RA et al. The genome sequence of Drosophila melanogaster. Science 2000; 287:2185-2195.
3. Handler AM, Gomez SP, O'Brochta DA. A functional analysis of the P-element gene-transfer vector in insects. Arch Insect Biochem Physiol 1993; 22:373-84.
4. Sarkar A, Coates CJ, Whyard S et al. The Hermes element from Musca domestica can transpose in four families of cyclorrhaphan flies. Genetica 1997; 99:15-29.
5. Loukeris TG, Livadaras I, Arca B et al. Gene transfer into the medfly, Ceratitis capitata, with a Drosophila hydei transposable element. Science 1995; 270:2002-5.
6. Jasinskiene N, Coates CJ, Benedict MQ et al. Stable transformation of the yellow fever mosquito, Aedes aegypti, with the Hermes element from the housefly. Proc Natl Acad Sci USA 1998; 95:3743-7.
7. Chalfie M, Tu Y, Euskirchen G et al. Green fluorescent protein as a marker for gene expression. Science 1994; 263:802-5.
8. Kokoza V, Ahmed A, Wimmer EA et al. Efficient transformation of the yellow fever mosquito Aedes aegypti using the piggyBac transposable element vector pBac[3xP3-EGFP afm]. Insect Biochem Mol Biol 2001; 31:1137-43.
9. Coates CJ, Jasinskiene N, Morgan D et al. Purified mariner (Mos1) transposase catalyzes the integration of marked elements into the germ-line of the yellow fever mosquito, Aedes aegypti. Insect Biochem Mol Biol 2000; 30:1003-8.
10. Kapetanaki MG, Loukeris TG, Livadaras I et al. High frequencies of Minos transposon mobilization are obtained in insects by using in vitro synthesized mRNA as a source of transposase. Nucleic Acids Res 2002; 30:3333-40.
11. Coates CJ, Jasinskiene N, Miyashiro L et al. Mariner transposition and transformation of the yellow fever mosquito, Aedes aegypti. Proc Natl Acad Sci USA 1998; 95:3748-51.
12. Rowan KH, Orsetti J, Atkinson PW et al. Tn5 as an insect gene vector. Insect Biochem Mol Biol 2004; 34:695-705.
13. Allen ML, O'Brochta DA, Atkinson PW et al. Stable, germ-line transformation of Culex quinquefasciatus (Diptera: Culicidae). J Med Entomol 2001; 38:701-10.
14. Catteruccia F, Nolan T, Loukeris TG et al. Stable germline transformation of the malaria mosquito Anopheles stephensi. Nature 2000; 405:959-62.
15. Nolan T, Bower TM, Brown AE et al. Piggybac-mediated germline transformation of the malaria mosquito Anopheles stephensi using the red fluorescent protein dsRED as a selectable marker. J Biol Chem 2002; 277:8759-62.
16. Grossman GL, Rafferty CS, Clayton JR et al. Germline transformation of the malaria vector, Anopheles gambiae, with the piggyBac transposable element. Insect Mol Biol 2001; 10:597-604.
17. Perera OP, Harrell IInd RA, Handler AM. Germ-line transformation of the South American malaria vector, Anopheles albimanus, with a piggyBac/EGFP transposon vector is routine and highly efficient. Insect Mol Biol 2002; 11:291-7.
18. Ryder E, Russell S. Transposable elements as tools for genomics and genetics in Drosophila. Brief Funct Genomic Proteomic 2003; 2:57-71.
19. Wilson R, Orsetti J, Klocko AD et al. Post-integration behavior of a Mos1 mariner gene vector in Aedes aegypti. Insect Biochem Mol Biol 2003; 33:853-63.
20. O'Brochta DA, Sethuraman N, Wilson R et al. Gene vector and transposable element behavior in mosquitoes. J Exp Biol 2003; 206:3823-34.

21. Jasinskiene N, Coates CJ, Ashikyan A et al. High efficiency, site-specific excision of a marker gene by the phage P1 creloxP system in the yellow fever mosquito, Aedes aegypti. Nucleic Acids Res 2003; 31:e147.
22. Lycett GJ, Kafatos FC, Loukeris TG. Conditional expression in the malaria mosquito Anopheles stephensi with Tet-On and Tet-Off systems. Genetics 2004; 167:1781-90.
23. Brown AE, Bugeon L, Crisanti A et al. Stable and heritable gene silencing in the malaria vector Anopheles stephensi. Nucleic Acids Res 2003; 31:e85.
24. Coates CJ, Jasinskiene N, Pott GB et al. Promoter-directed expression of recombinant firefly luciferase in the salivary glands of Hermes-transformed Aedes aegypti. Gene 1999; 226:317-25.
25. Moreira LA, Edwards MJ, Adhami F et al. Robust gut-specific gene expression in transgenic Aedes aegypti mosquitoes. Proc Natl Acad Sci USA 2000; 97:10895-8.
26. Allen ML, Christensen BM. Flight muscle-specific expression of act88F: GFP in transgenic Culex quinquefasciatus Say (Diptera: Culicidae). Parasitol Int 2004; 53:307-14.
27. Kim W, Koo H, Richman AM et al. Ectopic expression of a cecropin transgene in the human malaria vector mosquito Anopheles gambiae (Diptera: Culicidae): Effects on susceptibility to Plasmodium. J Med Entomol 2004; 41:447-55.
28. Shin SW, Kokoza VA, Raikhel AS. Transgenesis and reverse genetics of mosquito innate immunity. J Exp Biol 2003; 206:3835-43.
29. Moreira LA, Ito J, Ghosh A et al. Bee venom phospholipase inhibits malaria parasite development in transgenic mosquitoes. J Biol Chem 2002; 277:40839-43.
30. Ito J, Ghosh A, Moreira LA et al. Transgenic anopheline mosquitoes impaired in transmission of a malaria parasite. Nature 2002; 417:452-5.
31. James AA. Blocking malaria parasite invasion of mosquito salivary glands. J Exp Biol 2003; 206:3817-21.
32. Travanty EA, Adelman ZN, Franz AW et al. Using RNA interference to develop dengue virus resistance in genetically modified Aedes aegypti. Insect Biochem Mol Biol 2004; 34:607-13.
33. Guimond N, Bideshi DK, Pinkerton AC et al. Patterns of Hermes transposition in Drosophila melanogaster. Mol Genet Genomics 2003; 268:779-90.
34. Hacker U, Nystedt S, Barmchi MP et al. Piggybac-based insertional mutagenesis in the presence of stably integrated P elements in Drosophila. Proc Natl Acad Sci USA 2003; 100:7720-5.
35. Lorenzen MD, Berghammer AJ, Brown SJ et al. Piggybac-mediated germline transformation in the beetle Tribolium castaneum. Insect Mol Biol 2003; 12:433-40.
36. Catteruccia F, Godfray HC, Crisanti A. Impact of genetic manipulation on the fitness of Anopheles stephensi mosquitoes. Science 2003; 299:1225-7.
37. Irvin N, Hoddle MS, O'Brochta DA et al. Assessing fitness costs for transgenic Aedes aegypti expressing the GFP marker and transposase genes. Proc Natl Acad Sci USA 2004; 101:891-6.
38. Moreira LA, Wang J, Collins FH et al. Fitness of anopheline mosquitoes expressing transgenes that inhibit Plasmodium development. Genetics 2004; 166:1337-41.
39. Mori A, Chadee DD, Graham DH et al. Reinvestigation of an endogenous meiotic drive system in the mosquito, Aedes aegypti (Diptera: Culicidae). J Med Entomol 2004; 41:1027-33.
40. Sinkins SP. Wolbachia and cytoplasmic incompatibility in mosquitoes. Insect Biochem Mol Biol 2004; 34:723-9.
41. Sinkins SP, Hastings IM. Male-specific insecticide resistance and mosquito transgene dispersal. Trends Parasitol 2004; 20:413-6.
42. Adelman ZN, Jasinskiene N, Vally KJ et al. Formation and loss of large, unstable tandem arrays of the piggyBac transposable element in the yellow fever mosquito, Aedes aegypti. Transgenic Res 2004; 13:411-25.

CHAPTER 6

Schistosome Long Terminal Repeat Retrotransposons

Claudia S. Copeland, Thewarach Laha and Paul J. Brindley*

Abstract

The human schistosomes, blood flukes of the Genus *Schistosoma*, have a large genome estimated to be at least 270 megabase pairs (haploid) in size, arrayed on eight pairs (2n = 16) of chromosomes including the Z (male) and W (female) sex chromosomes. The genome appears to include about 14,000 protein encoding genes and a large amount of repetitive sequences. Much of this repetitive component of the schistosome genome is comprised of mobile genetic elements. Of these elements, an expanding number of discrete long terminal repeat (LTR) retrotransposons have been identified in the past five years. The identity, structure, phylogenetic relationships, and size contributions of these schistosome LTR retrotransposons are reviewed here. These elements include the *gypsy*-like or *Pao/BEL* -like retrotransposons *Boudicca, Sinbad, fugitive, Saci-1, Saci-2, and Saci-3* from *Schistosoma mansoni* and *Gulliver* from *Schistosoma japonicum*.

Introduction

Schistosomes and Schistosomiasis

Schistosomiasis is considered the most important of the human helminthiases, and second only to malaria among tropical parasitic diseases, in terms of morbidity.[1,2] Up to 200 million people are estimated to be infected worldwide, with annual deaths as high as 200,000.[2] Symptoms range from chronic pain due to granulomatous lesions of the liver and intestines to hepatosplenomegaly, portal hypertension, ascites, diarrhea, and impaired physical and cognitive development in children.[3,4] Female genital schistosomiasis, a common condition among women infected with *Schistosoma haematobium*, can lead to abortion or premature delivery as well as lesions throughout the reproductive organs, and can facilitate sexually transmitted disease infection.[5] Moreover, interactions with other infectious diseases can induce increased pathology, as with coinfection with hepatitis C, in which liver damage can be more severe than in patients with either disease alone[6] and with HIV, where viral load has been found to be significantly higher in patients coinfected with helminth parasites including schistosomes,[5] presumably due to rapid initial viral replication facilitated by the immune response to the parasites. Furthermore, schistosomiasis is associated with bladder[7] and other cancers.[8]

The life cycle of *Schistosoma* involves parasitism of both vertebrates and snail species specific for each species of schistosome. Infectious larvae known as cercariae emerge from the snails into a body of water, where they initiate infection by direct penetration of human skin. In the

*Corresponding Author: Paul J. Brindley—Department of Microbiology, Immunology and Tropical Medicine, George Washington University Medical Center, Ross Hall, Room 448, 2300 Eye Street, NW, Washington, DC 20037 USA. Email: mtmpjb@gwumc.edu

Mobile Genetic Elements in Metazoan Parasites, edited by Paul J. Brindley.

human host, the worms develop into male and female adults which live together within the mesenteric venules of the intestines, and release eggs into the blood stream. To perpetuate the life cycle, the eggs exit from the blood vessels and traverse the intestinal wall, facilitated by secreted proteolytic enzymes and their spines, and pass out in the feces to fresh water. Though chemotherapy is available, its effectiveness is limited by continuous reinfection upon subsequent exposure to water containing cercariae. Furthermore, symptoms do not necessarily resolve upon chemotherapeutic cure of the infection, and chronic symptoms of the disease can remain with the patient for life. No vaccine is currently available.

Health education in combination with drug therapy forms the cornerstone of the World Health Organization's strategy to combat schistosomiasis. This strategy has met with mixed success. Though the endemic distribution of schistosomiasis has changed in the past 50 years, overall, the estimated number of infected persons and those at risk of infection has not been reduced.[9-11] It is hoped that an enhanced understanding of the schistosome genome can be expected to lead to long-term strategies for the control of schistosomiasis.

The Schistosome Genome

Schistosome species infecting humans include *Schistosoma mansoni*, *Schistosoma japonicum*, *Schistosoma haematobium*, and, to a lesser extent, *Schistosoma intercalatum* and *Schistosoma mekongi*.[12] Of these, the African blood fluke, *S. mansoni*, and the Asian blood fluke, *S. japonicum* have been characterized reasonably well at the genomic level. Schistosomes have comparatively large genomes, estimated at ~270 megabase pairs for the haploid genome of *Schistosoma mansoni*, arrayed on seven pairs of autosomes and one pair of sex chromosomes.[13] For comparison, the schistosome genome is about the same size as that of the puffer fish, *Takifugu rubripes*, two to three times the size of that of the angiosperm, *Arabidopsis thaliana*, or the free-living nematode, *Caenorhabditis elegans*, ten times the size of the *Plasmodium falciparum* genome, and about one tenth the size of the human genome. The other major schistosome species parasitizing humans probably have a genome of similar size to that of *S. mansoni*, based on the similarity in appearance of their karyotypes,[14] though very recent estimates based on genome sequencing indicate that the genome of *S. japonicum* may be as large as 400 megabase pairs or more (Wang S-Y, personal communication). Although no schistosome genome has been sequenced in its entirety, several hundred thousand schistosome expressed sequence tags (ESTs) and genome survey sequences have been deposited in GenBank, and these may represent the complete transcriptomes of *S. mansoni* and *S. japonicum*, covering ~14,000 genes in *S. mansoni* and ~13,000 genes in *S. japonicum*.[15] These ESTs appear to represent most, if not all, *S. mansoni* and *S. haematobium* genes, with expressed gene content of schistosomes previously estimated to be 15,000-20,000.[17] The mobile genetic elements (MGEs) of the schistosome genome include transposons, which replicate via a "cut and paste" process of direct transposition as DNA, retrotransposons, which replicate through an RNA mediated process, and SINE-like elements, which use the enzymes of retrotransposons to replicate (Fig. 1). Recent findings suggest that up to half of the entire schistosome genome may be comprised of repetitive sequences, with much of this repetitive complement comprised of mobile genetic elements.[18] Mobile genetic elements appear to make up at least one quarter of the schistosome genome. Here we review the LTR retrotransposons from the genomes of schistosome species. The identity, structure, phylogenetic relationships, and contribution of these elements to genome size in schistosomes are described, and we address their probable role in schistosome evolution and potential utility in applications such as introducing transgenes into schistosomes.

LTR Retrotransposons

The retrotransposon component of the schistosome genome includes both the retroviral-like LTR retrotransposons, so named because they are framed by long terminal direct repeats, and non-LTR retrotransposons, which often include A-rich 3' termini in their genomic forms. LTR retrotransposons resemble retroviruses in their structure and intracellular life cycles (Figs. 2, 3).

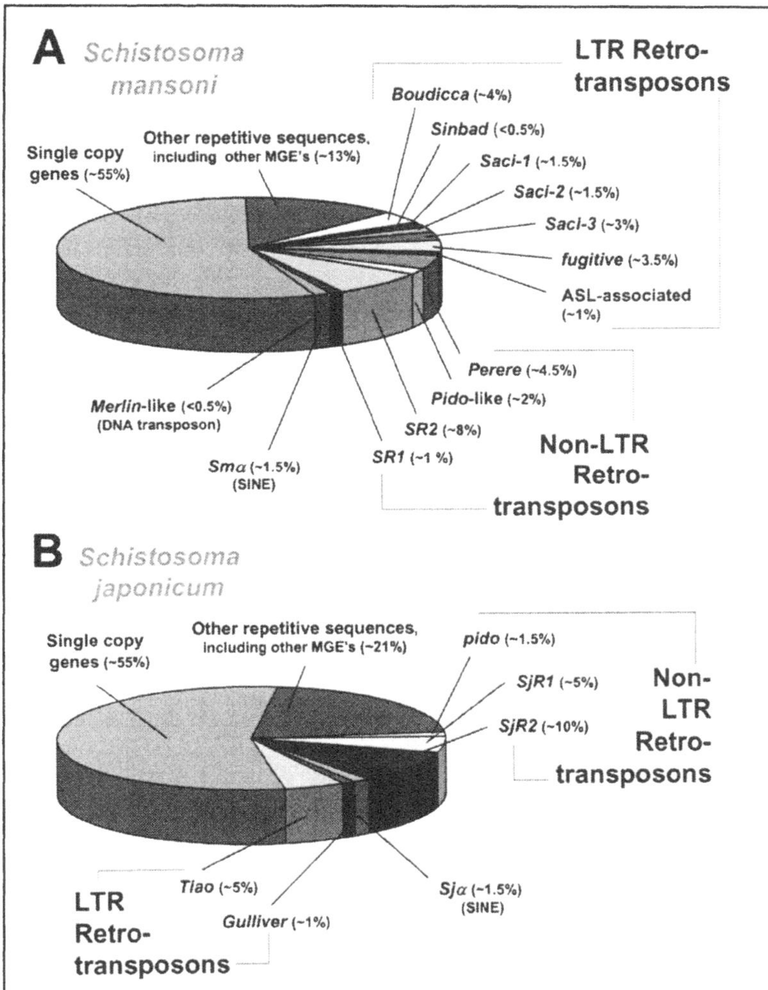

Figure 1. Contribution of mobile genetic elements in the genomes of (A) *Schistosoma mansoni* and (B) *Schistosoma japonicum* expressed in pie chart format. Predicted approximate contributions (in terms of simple DNA mass) of various mobile genetic elements, deducted from published and gene database sources are illustrated as pie slices, with percentages in text following the name of each mobile genetic element or genome category.

These elements are typically 5-10 kb in length. Their general structure consists of two open reading frames (ORFs) flanked by long direct terminal repeats of ~200-600 bp in length (Fig. 2). Some, such as *gypsy* from *Drosophila melanogaster*, *Osvaldo* from *Drosophila buzzatii*, and *Tas* from *Ascaris lumbricoides*, include a third ORF, *env*, encoding the envelope glycoprotein characteristic of retroviruses. Retroviruses probably evolved from LTR retrotransposons, mediated by the acquisition of envelope proteins that facilitated extracellular existence and horizontal transmission between cells and species.[19,20] The LTRs play a pivotal role in transposition and the initiation of transcription, with the 5' LTR encoding the promoter. The first ORF, *gag*, encodes a polyprotein precursor that is later processed to yield the structural proteins making up the virion core (Fig. 3). Of these, the nucleocapsid protein associates directly with the RNA, and

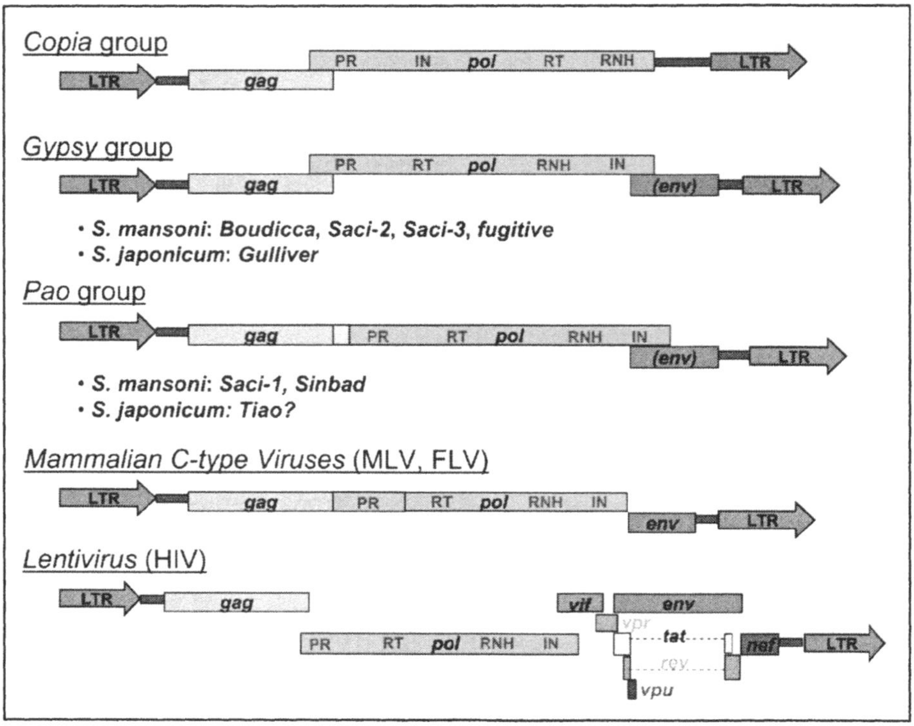

Figure 2. Diagrammatic representation of the genome structures of the three major LTR retrotransposon groups, *Copia/Ty1*, *Gypsy/Ty3* and *Pao/BEL*, and two related retroviruses, the Mammalian C-type Viruses and the Lentiviruses. The LTR retrotransposons so far characterized from the *S. mansoni* and *S. japonicum* genomes fall into the *Gypsy/Ty3* and *Pao/BEL* groups, and are listed below the structures characteristic of these groups. Abbreviations: LTR, long terminal repeat; *gag*, group associated antigen; *pol*, polyprotein; env, envelope; PR, protease, RT, reverse transcriptase, RH, RNaseH, IN, integrase. The HIV genes *vif, vpr, rev, vpu,* and *nef* are accessory proteins specific to the more complex replication cycle of HIV.

exhibits a characteristic cysteine/histidine motif, which appears to function as a zinc finger domain. Transcription of the second ORF, *pol*, is initiated by read-through or ribosome slippage from the *gag* transcript. *Pol* encodes the enzymes of the retrotransposon, and, like *gag*, is cleaved by protease to yield functional proteins. These enzymes include protease, which cleaves the polyprotein, reverse transcriptase, which is responsible for transcribing the RNA genome into double stranded DNA, RNAse H, which removes the original RNA template from the newly synthesized DNA strand, allowing the complementary DNA strand to form, and integrase, which inserts the newly replicated proviral form of the retrotransposon into the host genome (Fig. 3). In retroviruses and LTR retrotransposons with an *env* gene, the envelope protein associates with the cell membrane, which envelops the virion core, allowing the viral particle to bud off from the host cell (Fig. 3). Envelope facilitates infection via attachment to specific cell surface receptors. Thus, in addition to vertical transmission in the germ line, LTR retrotransposons with *envelope* genes are capable of extracellular existence and horizontal transmission.

The two best understood retrotransposon families, *Copia/Ty1* and *gypsy/Ty3*, can be differentiated on the basis of their *pol* domain order, with a domain order of PR-RT-RNAse H-IN for the *gypsy/Ty3* retrotransposons and PR-IN-RT-RNAse H for the *Copia/Ty1* retrotransposons (Fig. 2). In addition to the *gypsy/Ty3* and *Copia/Ty1* families, two other

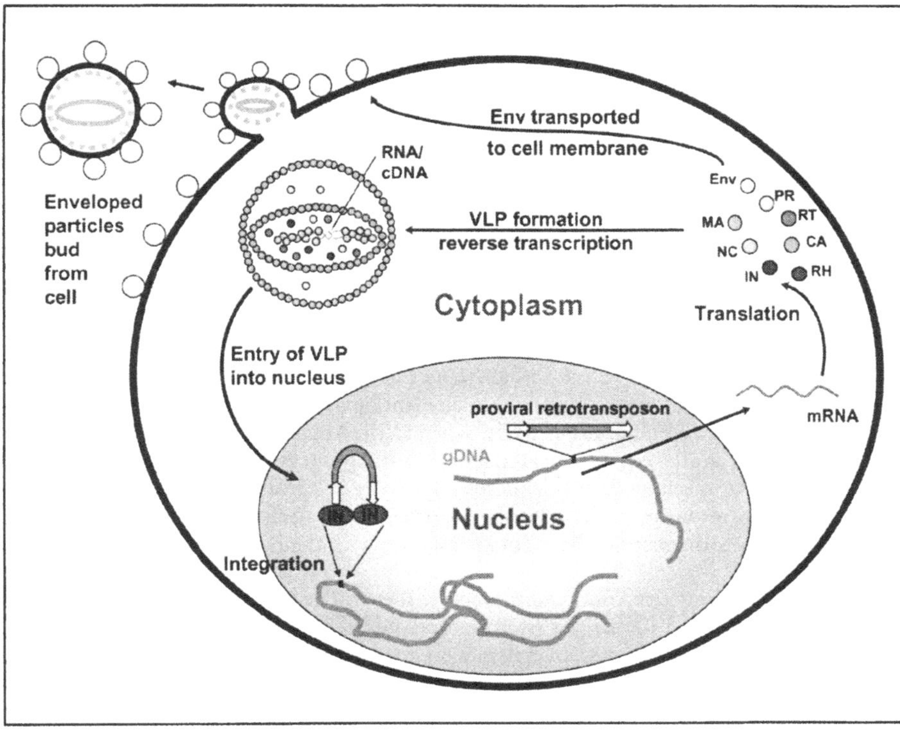

Figure 3. Diagrammatic representation of the replication cycle of an LTR retrotransposon. The proteins making up the virion core, NC, CA, and MA, are encoded by *gag*, whereas the enzymes PR, RT, RH, and IN are encoded by *pol*. *Env*, if present, is produced by splicing and is not a part of *gag* or *pol*. The VLP of LTR-retrotransposons is analogous to the virion of retroviruses. Abbreviations: gDNA, genomic DNA; mRNA, messenger RNA; cDNA, complementary DNA; NC; nucleocapsid; CA, capsid; MA, matrix; env, envelope; PR, protease, RT, reverse transcriptase, RH, RNaseH, IN, integrase; VLP, Viral-Like Particle. The illustration was adapted from Havecker et al[67] (with permission), using structural information from Coffin et al.[68]

families of LTR-retrotransposons are present in the schistosome genome that do not fit into either of these clades. These include the *Pao* group retrotransposons, which have the same structural domain order as the *gypsy/Ty3* family, but form a distinct group at the primary sequence level, and the *Penelope* group of retrotransposons. The long terminal repeats of the *Penelope* retrotransposons, known as PLTRs, are distinctive in structure from the LTRs of LTR-retrotransposons, and indeed the *Penelope* elements were initially designated as a novel branch of non-LTR retroelements.[21] The open reading frames of the element extend into each PLTR, and the 5' PLTR is often preceded by an inverted PLTR.[21,22] The presence of introns in some members of this group, indicating the ability to replicate using an RNA-independent pathway, along with several other unique features indicate that these elements are distinct from both LTR and non-LTR retrotransposons.

Contributions of Transposable Elements to Genome Size, Structure and Evolution

The comparatively large size of the genome of *Schistosoma* is to a large extent due to the presence of numerous mobile genetic elements (Fig. 1). In addition to their contribution to genome size, retrotransposons also serve as a source of genetic variation by numerous means.[23]

Retrotransposons appear to be quite active in schistosomes, with transposable element ESTs making up from 3% (adult worms) to 14% (cercariae) of the transcriptome in *S. mansoni.*[24] The movement of retrotransposons can lead to insertional mutagenesis if they integrate into the coding portions of genes (see, for example, Casacuberta and Santiago).[25] Insertion into noncoding regions can also bring about significant changes. Integration into promoters or other regulatory elements can silence or enhance gene expression through disruption of these regulatory elements, and insertion into introns can lead to alternative splicing, resulting in exon skipping and truncated proteins.[26,27] Also, the LTRs themselves include transcriptional promoters, regulators, and terminators that can influence the transcription of a gene if the retrotransposon integrates near the gene.[25] Another epigenetic phenomenon that can come into play when transposable elements insert near genes is the spread of gene silencing from the element to the nearby gene. This type of gene silencing due to transposon-induced RNA interference (RNAi) has been observed in *Drosophila.*[28] Interestingly, transposable element activity seems to increase in the life stages of *S. mansoni* related to entry of the human host: transposable elements make up 14% and 7.4% of ESTs of cercariae and schistosomula, respectively, whereas they make up 3.0%, 3.7%, 4.9%, and 3.0% of ESTs of adults, eggs, miracidia, and mother sporocysts (germballs), respectively.[24] It is intriguing to speculate that this increased activity in cercariae and schistosomula may have evolved in response to the advantage conferred on a schistosome population by an increase in genetic diversity at the point of interface with the hostile immune system of the human host.

Genes can also be derived from transposable elements. The human genome contains dozens of genes that appear to be derived from transposable elements,[29] and *Arabidopsis* has been shown to have derived both promoter elements and exons from transposable elements.[30]

Finally, multi-copy elements can influence genomic structure through homologous recombination, resulting in chromosomal rearrangements such as deletions, duplications, or reciprocal translocations.[31,32] One example of a large scale, unexplained deletion in the schistosome genome is found within the cathepsin D gene of *S. mansoni.* The *S. mansoni* cathepsin D gene is homologous at the primary sequence level and has a similar exon-intron boundary structure to that the gene encoding cathepsin D in vertebrates, with one notable exception: two large exons appear to have undergone complete deletion.[33] The introns of this schistosome gene are large and highly populated by retrotransposons and retrotransposon fragments, which suggested to Morales et al[33] that homologous recombination between retrotransposon copies in two introns may have facilitated deletion of the exon between them.

Transposable elements, then, are capable of inducing mutagenesis and, as a result, variability in a population. Indeed, it has been suggested by Barbara McClintock and others that genomes have evolved to activate transposable elements in response to environmental stress.[34] Whether or not genomes have evolved mechanisms to specifically harness the power of transposable element-induced mutation, with such a range of influence on genome change and variability, transposable elements clearly play a major role in genome evolution.

Schistosome Retrotransposable Elements: Gypsy/Ty3 Retrotransposons

Boudicca

Boudicca is a ~6 kb long terminal repeat retrotransposon from *S. mansoni*[35] (5,858 bp long for the original, well-characterized 53-J-5 copy of the retrotransposon obtained from a bacterial artificial chromosome [BAC] library of the *S. mansoni* genome.[36] Two 328bp LTRs flank a coding region consisting of open reading frames representing the *gag* and *pol* polyproteins, 5' and 3' untranslated regions, and, at least in the BAC 53-J-5 copy, an additional unknown third open reading frame (Fig. 2). *Boudicca* is a high copy number element, estimated at 2,000 to 3,000 copies per haploid genome, and is actively transcribed in adult worms, cercariae, and

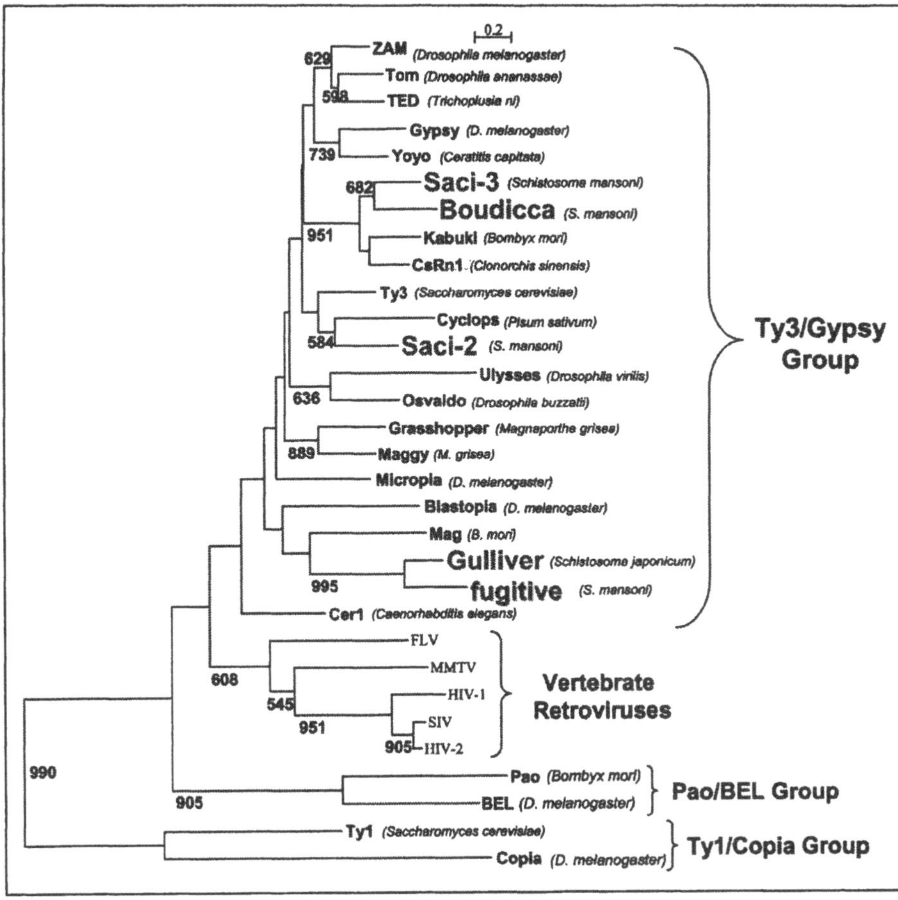

Figure 4. Phylogeny of *Gulliver, Boudicca, Saci-2, Saci-3,* the *fugitive,* and related elements. Phylogenetic tree comparing the relationships among the schistosome *gypsy* group LTR retrotransposons *Gulliver, Boudicca, Saci-2, Saci-3,* the *fugitive* and other LTR retrotransposons and retroviruses. Reverse transcriptase regions were aligned and tree files were generated using MacVector (Accelrys) and ClustalX software,[69] and trees were constructed using njplot software.[70] Gaps were excluded for the analysis, and a Clustal distance correction was employed to correct for multiple substitutions. Accession numbers: *BEL*: U23420, *Blastopia*: CAA81643, *Boudicca*: AY662653, *Cer1*: AAA50456, *Copia*: OFFFCP, *CsRn1*: AAK07486, *Cyclops*: AAL06415, HIV-1: P04585, HIV-2: AAA76841, FLV: NP_047255, *Fugitive*: BK005226, *Grasshopper*: AAA21442, *Gulliver*: AF243513, *Gypsy*: GNFFG1, *Kabuki*: BAA92689, *Mag*: S08405, *Maggy*: AAA33420, *Micropia*: S02021, MMTV: GNMVMM, *Osvaldo*: CAB39733, *Pao*: composite constructed from S33901, AB042118, and AB042119 (see Copeland et al.42), *Saci-2*: BK004069, *Saci-3*: BK004070, SIV: AAA47606, *TED*: AAA92249, *Tom*: CAA80824, *Ty1*: P47100, *Ty3*: AAA35184, *Ulysses*: CAA39967, *Yoyo*: T43046, *Zam*: CAA04050.

sporocysts. One of several characterized members of the *Kabuki/CsRn1* clade of *gypsy*-type LTR retrotransposons, *Boudicca*'s closest nonschistosome relatives are *PwRn1* from the lungworm *Paragonimus westermani*, *CsRn1* from the oriental liver fluke, *Clonorchis sinensis*, and *Kabuki* from the silkworm, *Bombyx mori*. This clade, closely related to the errantiviruses, is differentiated not only at the primary sequence level of reverse transcriptase (Fig. 4) but also by a unique Cys-His box structure, CHCC instead of the more common retroviral CCHC Cys-His box.[37]

Models of the secondary structure of the *Boudicca* transcript suggested that the first AUG was occluded by a stem loop structure, which in turn suggested a method of regulation of expression, at the level of translation, of *Boudicca* proteins.[38]

Fugitive

The *fugitive* element of *S. mansoni* belongs to the *mag*-like family of the *gypsy-Ty3* clade of long terminal repeat (LTR) retrotransposons.[39] It is closely related to the *mag*-like retrotransposon *Gulliver* from *S. japonicum*, but is dissimilar to several other LTR retrotransposons known from *S. mansoni* including *Boudicca, Saci-1, Saci-2, Saci-3,* and *Sinbad.* The full length *fugitive* element is 4,811 bp constituted of a single read-through open reading frame (ORF) of 4134 bp flanked at both ends by identical long terminal repeat sequences of 273 bp. This single ORF encodes retroviral-like *gag* and *pol* polyproteins, with a pol domain order of protease, reverse transcriptase, RNaseH and integrase. Gag of *fugitive* exhibits the distinctive double Cys-His motif characteristic of the *mag*-like retrotransposons.[40] Examination of schistosome transcriptome sequences in the public domain revealed that the *fugitive* was transcribed in at least six developmental stages of *S. mansoni*, while bioinformatics approaches and Southern hybridization analysis indicated that as many as 2000 copies of the *fugitive* were interspersed throughout the *S. mansoni* genome. Interestingly, the sequence of the first characterized copy (from BAC clone 49-J-14; accession no. BK005226) does not include any obvious mutations (stop or frame shift mutations). This apparent integrity, together with the identical sequences of the LTRs, suggests that the sequenced copy may be an active copy of this retrotransposon.

Gulliver

The other *mag*-family retrotransposon reported to date from schistosomes is *Gulliver*, from *Schistosoma japonicum. Gulliver* is ~4.8 kb in length and is present in multiple copies in the genome of *S. japonicum*[41] (Fig. 1). The LTRs of *Gulliver* are 259 bp in length and include RNA polymerase II promoter sequences, a CAAT signal and a TATA box. *Gulliver* exhibits features characteristic of a functional LTR retrotransposon including two read through (termination) ORFs encoding retroviral gag and pol proteins of 312 and 1,071 amino acid residues, respectively. The *gag* ORF encodes motifs conserved in nucleic acid binding proteins, while the *pol* ORF encodes conserved domains of aspartic protease, reverse transcriptase (RT), RNaseH and integrase, in that order, a pol pattern conserved in the *gypsy* lineage of LTR retrotransposons. *Gulliver*'s closest nonschistosome relatives include *mag* from *B. mori* and *Blastopia* from *D. melanogaster* (Fig. 4). Southern hybridization analysis indicated the presence of *Gulliver*-like elements in the genome of *S. mansoni*,[41] and these may be identical to the *fugitive* retrotransposon reported subsequently from *S. mansoni.*[39]

Saci-2

The *Saci-2* element of *S. mansoni* is 4,946 bp long and encodes a single ORF.[24] It is closely related to *Ty3* from *Saccharomyces cerevisiae. Saci-2* exhibits an unusual Cys-His box motif, with a lysine replacing a normally conserved histidine to yield a Cys-His box of the structure C(X2)C(X4)K(X4)C rather than the more conserved C(X2)C(X4)H(X4)C. This lysine replacement was conserved in 22 of 24 ESTs covering that region of the retrotransposon, and neither of the other two ESTs had a histidine in the ninth position, indicating that this replacement is not simply a mutation unique to a single *Saci-2* copy. *Saci-2* is transcribed in all six life stages of *S. mansoni.*[24]

Saci-3

Saci-3 is 5,217 bp long and is the second member of the *Kabuki/CsRn1* family to be found in the genome of *S. mansoni.*[24] *Saci-3* is closely related to *Boudicca* in primary sequence structure, and also has three open reading frames. This similarity, combined with low bootstrap values for separation of their RT amino acid sequences, call strict differentiation of *Boudicca* and *Saci-3* into question and raise the possibility that they are in fact distantly diverged copies of the same retrotransposon. However, the LTRs of these two retrotransposons are dissimilar as determined by phylogenetic analysis (bootstrap value for their separation is 96%), and their lengths are different (~210-290bp for *Saci-3* vs. ~330 bp for *Boudicca*), supporting the classification of these elements as closely related but distinct. Like *Boudicca*, *Saci-3* also shares the CHCC Cys-His box structure unique to the *Kabuki/CsRn1* clade and is transcribed in all six life stages of *S. mansoni*.

Copia/Ty1 Retrotransposons

No *copia*-like retrotransposons have been described (yet) in schistosomes. However, preliminary sequence data from the *S. mansoni* genome sequencing project of The Institute for Genomic Research (TIGR) reveals three *S. mansoni* gDNA BAC clones exhibiting significant ($p < 1$) similarity to the complete DNA sequence of *Copia* from *Drosophila melanogaster*. As of February 2005, these BAC sequences, 48-C-10, 31-L-12, and 49-A-11, were still classified as "preliminary sequence data" and so were not yet available in the public databases, but they are available from the TIGR database, http://www.tigr.org/tdb/e2k1/sma1/gene.shtml.

Pao/BEL Retrotransposons

Saci-1

The first *Pao*-like element to be found in the genome of *S. mansoni*, *Saci-1* is 5980bp in length, with LTRs of 848bp in length. *Saci-1* has a single ORF encoding a polyprotein 1680 amino acids long, including domains for protease, reverse transcriptase, RNAse H, and integrase, as well as a triple Cys-His box.[24] Cys-His boxes exhibiting three separate zinc finger domains are unique to and characteristic of *Pao/BEL* elements. *Saci-1* was originally described from ESTs found in six life stages of *S. mansoni*, indicating that this element is actively transcribed throughout the *S. mansoni* life cycle.

Sinbad

Sinbad, the second *Pao*-like element to be found in the *S. mansoni* genome, is 6,287 bp long and is flanked at its 5'- and 3'-ends by identical LTRs of 386 bp.[42] Like *Saci-1*, *Sinbad* displays a triple Cys-His RNA binding motif characteristic of *gag* genes of *Pao/BEL*-like elements,[43,44] followed by the enzymatic domains of protease, reverse transcriptase (RT), RNAseH, and integrase, in that order. Though their vastly different LTR lengths confirm that *Saci-1* and *Sinbad* are different elements, they are nevertheless closely related, as determined by phylogenetic analysis. Together with an uncharacterized element from the *Danio rerio* genome, *Sinbad* and *Saci-1* appear to make up a separate subfamily of *Pao/BEL* elements (Fig. 5). Southern hybridization and bioinformatics analyses indicated that there were about 50 copies of *Sinbad* in the *S. mansoni* genome. The presence of ESTs spanning *Sinbad* in six life stages of *S. mansoni*, along with the identical 5'- and 3'-LTR sequences, indicate that *Sinbad* is actively transcribed throughout the *S. mansoni* life cycle.

Tiao

Tiao, the first LTR type retrotransposon to be reported from any of the human schistosomes, is a high copy number (about 10,000) LTR retrotransposon found in *S. japonicum* (GenBank AF073334).[45] *Tiao* is related to the *Pao*-like retrotransposons *Kamikaze*, *Yamato*, and *Pao* from *B. mori*, and *Ninja*, *BEL*, and *Max* from *Drosophila melanogaster*. More sequence data are needed to determine more precisely where *Tiao* falls within the phylogeny of *Pao/BEL* elements.

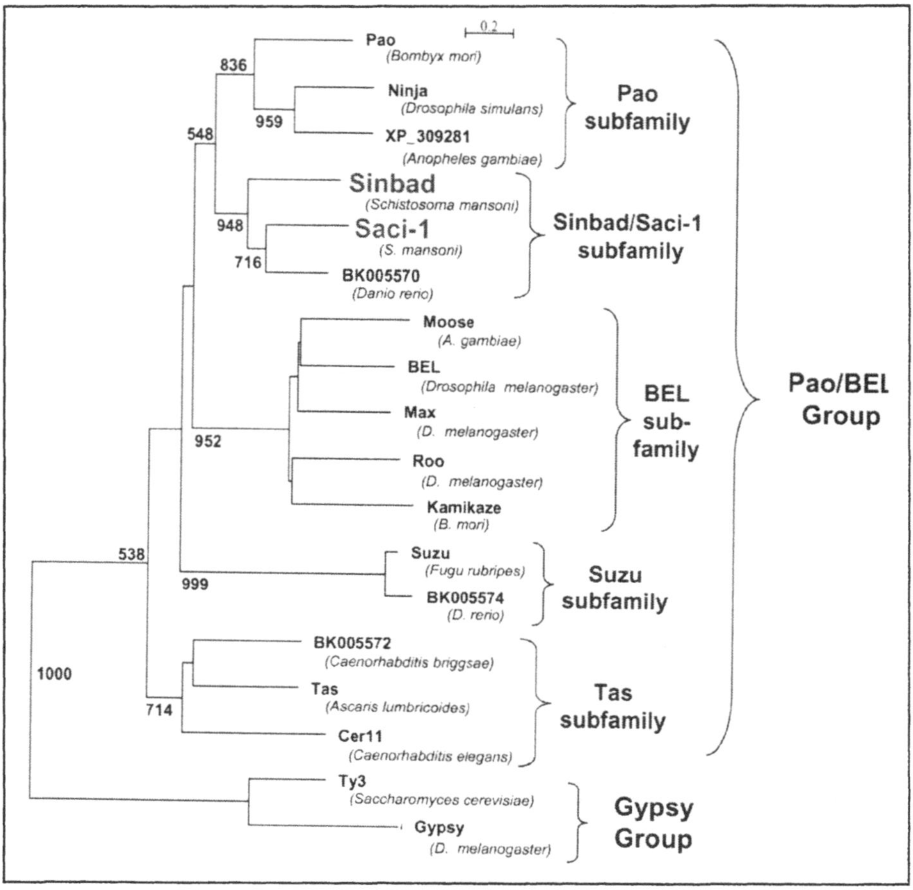

Figure 5. Phylogenetic tree comparing the relationships among the schistosome LTR retrotransposons *Sinbad, Saci-1* and other *Pao/BEL* retrotransposons, with *Ty3* and *Gypsy* as the outgroup. Reverse transcriptase regions were aligned and tree files were created using MacVector (Accelrys) and ClustalX software,[69] and trees were created using njplot software.[70] Gaps were excluded for the analysis, and a Clustal distance correction was employed to correct for multiple substitutions. Accession numbers: *BEL*: U23420, *Cer11*: AAA82437, *Gypsy*: GNFFG1, *Kamikaze*: AB042120, *Max*: CAD32253, *Moose*: AF060859, *Ninja*: T31674, *Pao*: composite constructed from S33901, AB042118, and AB042119 (see Copeland et al[42]), *Roo*: AAN87269, *Saci-1*: DAA04498, *Sinbad*: AY506538, *Suzu*: AF537216, *Tas*: Z29712, *Ty3*: AAA35184.

Other Schistosome LTR-Retrotransposons

Cercyon

The *Penelope*-like retrotransposons, typified by *Penelope* of *Drosophila virilis* and *Xena* of *Takifugu rubripes* and *Tetraodon nigroviridis*, compose a highly divergent retrotransposon group distinct from both the LTR- and non-LTR retrotransposons. Though bounded by long terminal direct repeats, these elements lack Gag-, protease-, and RNAseH encoding regions, and appear to replicate in a manner more similar to that of the non-LTR retrotransposons, as indicated by an abundance of 5' truncated copies and variable target site duplication lengths similar to those of L1-type elements.[22,46] Unique to the *Penelope*-like elements is the phenomenon of open reading frames extending into both the 5' and 3' long terminal repeats, leading to their designation as "PLTRs" to distinguish them from the LTRs of *Gypsy*-like, *Copia*-like, and *Pao*-like retrotransposons.[22] Indicative of a distinct mechanism of replication is the presence of an inverted PLTR flanking the 5' direct PLTR in many members of this group.[21,22] The *Penelope* elements are the only known retroelements to contain introns,[46] a further indication of a unique mechanism of replication. The *S. mansoni* genome contains a *Penelope*-like retrotransposon, designated *Cercyon* (GenBank accession no. BK000685). *Cercyon* is 2370 bp long, encodes reverse transcriptase and endonuclease domains, and is closely related to an unnamed element from the whipworm, *Trichuris*.[46]

ASL-Associated and Undescribed LTR Retrotransposons

Within the locus of the *S. mansoni* gene encoding adenylosuccinate lyase, intron 6 contains two 500 bp direct LTRs flanking a 1.9 kb sequence with homology to LTR retrotransposon type *gag* and *pol*.[47] Though probably truncated, this was the first retrotransposon to be identified in *S. mansoni*. In addition, at least three other LTR type retrotransposons related to *Pao*, *Mag*, and *Osvaldo* are present in the genome of *S. mansoni*, awaiting characterization (Copeland and Brindley, unpublished).

Applications to the Study of Schistosomes

Genome Sequencing Issues—Shotgun Sequencing

The advent of shotgun sequencing has served to accelerate the process of genome sequencing. The speed and efficiency of this method, however, comes at the cost of uncertainty at the point of joining individual runs to produce large contigs.[48] Multi-copy elements, and particularly high copy number elements such as many LTR retrotransposons, pose a significant problem at the stage of construction of larger contigs from individual sequence runs. Multi-copy elements raise the spectre of contigs including one copy of the element being falsely joined to contigs including a different copy of the element, or a different element with a similar sequence, as with reverse transcriptase, which includes well-conserved domains.[49] Detailed understanding of these elements, including their sequences, their intercopy variability, and their copy number, can serve to alert analysts to "high risk" repeat regions and the need for further scrutiny in such regions, using for example methods such as those detailed by Tammi et al,[48] to ensure the quality of sequencing of the *Schistosoma* genomes by this method.

Development of Transgenesis/Mutagenesis Vectors

The ability to introduce transgenes into genomes of pathogenic organisms has revolutionized molecular approaches to microbial diseases. Whereas this revolution has included protozoan parasites, and indeed mosquito and other insect vectors, so far it has largely bypassed medical helminthology because neither cell lines nor transgenesis systems have been developed for parasitic worms. Consequently, molecular research in schistosomes and schistosomiasis, and indeed other helminthoses, has been structurally disadvantaged in comparison with other microbes.[50] Since MGEs are valued as vectors for genetic transformation of other invertebrate genomes,[51] likewise they hold promise for genetic transformation of schistosomes. The use of transposons, such as the *Merlin*-like transposon[52] of *S. mansoni*, holds potential as a transformation strategy, as does the use of retroviruses, which have demonstrated utility in some insect species.[53] Though vertebrate retroviruses such as Moloney murine leukemia virus (Mo-MLV) have been widely used for transgenesis studies, including human gene therapy research, a chief drawback with retroviruses is that they typically exhibit narrow host ranges. Though this can be mitigated by procedures such as pseudotyping, an ideal LTR-type transgenesis vector would be one that already integrates into the schistosome genome as part of the vector's natural life cycle. The active search for and characterization of endogenous LTR retrotransposons in schistosomes have already yielded several possible candidates (see Figs. 4,5). It seems feasible that schistosome retrotransposons such as these can be adapted directly as transgenesis vectors given similar success with other mobile elements.[54-56]

Insertional Mutagenesis

The goal of transgenesis leads to many other secondary objectives. One of these is the generation of collections of mutants, to enable examination of gene function through reverse genetics. Mobile genetic elements can be used to accomplish this goal. Transposable elements have been used for large scale insertional mutagenesis for the purposes of creating mutant assemblages of *Arabidopsis thaliana,*[57] rice,[58] *Drosophila melanogaster,*[59] *C. elegans*[60] and the zebrafish, *Danio rerio.*[61] The use of a marker gene flanked by LTRs of an endogenous schistosome LTR retrotransposon would be one of many ways to use these elements to deduce gene function through reverse genetics studies. An increased understanding of insertion sites and specificity/generality of individual schistosome LTR retrotransposons would facilitate the foundation for these kinds of studies of the schistosome genome.

Knockout of Cell-Cycle Regulatory Genes to Create a Cell Culture Line

Schistosome research has been inhibited by the lack of a cell culture line for any of the schistosome species. Stable cell culture lines from humans and other species have been created from cancer cells, which have themselves been naturally transformed through destruction of protective cell-cycle regulatory genes, such as p53 (e.g., ref. 62). If analogous cell-cycle regulatory genes were to be discovered in schistosomes, their disruption could lead efforts to establish stable schistosome cell lines. One approach to such disruption could be the use of retrotransposons for insertional mutagenesis.

Other Applications

In addition, retrotransposons and other MGEs can be of value in comparative genomics. For example, the difference in size between the *Caenorhabditis briggsae* (~104 MB) and *C. elegans* (100.3 MB) genomes is almost entirely due to repetitive sequence, which accounts for 22.4 % of the C. briggsae genome in contrast to 16.5 % of the *C. elegans* genome. Few, if any, repeat families are shared, suggesting that most were acquired after the two species diverged or are undergoing rapid evolution.[63] Further, MGEs could provide paleontological information about the evolution of the genes at their integration sites, for example, in relation to orthologous target genes in sister host taxa.[64,65] Due to their interspersed nature in genomes, and their ability to move within genomes, PCR based techniques targeting these kinds of elements can be employed for identification of insertion/deletion polymorphic markers, and/or of transgene integrations (e.g., ref. 66). Finally, MGEs should also be of value in epidemiological and clinical studies because the interspersed nature of these elements can produce genetic fingerprints, which might be associated with clinically or otherwise relevant traits.

Acknowledgements

This work was supported in part by a Tulane Graduate School dissertation fellowship to CSC and a Burroughs Wellcome Fund Scholar Award in Molecular Parasitology to PJB.

References

1. Crompton DWT. How much human helminthiasis is there in the world? J Parasitol 1999; 85:397-403.
2. Chitsulo L, Loverde P, Engels D. Schistosomiasis. Nature Reviews Microbiology 2004; 2:12-13.
3. Lambertucci JR. Schistosoma mansoni: Pathological and clinical aspects. In: Jordan P, Webbe G, Sturrock RF, eds. Human Schistosomiasis. Wallingford: CAB International, 1993:195-235.
4. Nokes C, McGarvey ST, Shiue L et al. Evidence for an improvement in cognitive function following treatment of Schistosoma japonicum infection in Chinese primary schoolchildren. Am J Trop Med Hyg 1999; 60:556-65.
5. Poggensee G, Feldmeier H. Female genital schistosomiasis: Facts and hypotheses. Acta Trop 2001; 79(3):193-210.

6. Gad A, Tanaka E, Orii K et al. Relationship between hepatitis C virus infection and schistosomal liver disease: Not simply an additive effect. J Gastroenterol 2001; 36(11):753-8.
7. Haitel A, Posch B, El-Baz M et al. Bilharzial related, organ confined, muscle invasive bladder cancer: Prognostic value of apoptosis markers, proliferation markers, p53, E-cadherin, epidermal growth factor receptor and c-erbB-2. J Urol 2001; 165(5):1481-7.
8. Qiu DC, Hubbard AE, Zhong B et al. A matched, case-control study of the association between Schistosoma japonicum and liver and colon cancers, in rural China. Ann Trop Med Parasitol 2005; 99(1):47-52.
9. Bergquist N. Schistosomiasis: From risk assessment to control. Trends Parasitol 2002; 18:309-314.
10. Chitsulo L, Engels D, Montresor A et al. The global status of schistosomiasis and its control. Acta Trop 2000; 77:41-51.
11. Zhou X, Chen M, McManus D et al. Schistosomiasis control in the 21st century. Proceedings of the International Symposium on Schistosomiasis, Shanghai, July 4-6, 2001. Acta Tropica 2002; 82:95-114.
12. Markell EK, John DT, Krotoski WA. Markell and voge's medical parasitology. 8th ed. W.B. Saunders Company, a division of Harcourt Brace & Company, 1999.
13. Simpson AJG, Sher A, McCutchan TF. The genome of Schistosoma mansoni: Isolation of DNA, its size, bases and repetitive sequences. Mol Biochem Parasitol 1982; 6:125-137.
14. Hirai H, Taguchi T, Saitoh Y et al. Chromosomal differentiation of the Schistosoma japonicum complex. Int J Parasitol 2000; 30(4):441-452.
15. Verjovski-Almeida S, DeMarco R, Martins EA et al. Transcriptome analysis of the acoelomate human parasite Schistosoma mansoni. Nat Genet 2003; 35:148-157.
16. Hu W, Yan Q, Shen D-K et al. Evolutionary and biomedical implications of a Schistosoma japonicum complementary DNA resource. Nat Genet 2003; 35:139-147.
17. Johnston DA, Blaxter ML, Degrave WM et al. Genomics and the biology of parasites. Bioessays 1999; 21(2):131-47.
18. Brindley PJ, Laha T, McManus DP et al. Mobile genetic elements colonizing the genomes of metazoan parasites. Trends Parasitol 2003; 19:79-87.
19. Temin HM. Origin of retroviruses from cellular moveable genetic elements. Cell 1980; 21:599-600.
20. Malik HS, Henikoff S, Eickbush TH. Poised for contagion: Evolutionary origins of the infectious abilities of invertebarate retroviruses. Genome Res 2000; 10:1307-1318.
21. Lyozin GT, Makarova KS, Velikodvorskaja VV et al. The structure and evolution of Penelope in the virilis species group of Drosophila: An ancient lineage of retroelements. J Mol Evol 2001; 52(5):445-56.
22. Dalle Nogare DE, Clark MS, Egar G et al. Xena, a full-length basal retroelement from tetraodontid fish. Mol Biol Evol 2002; 19:247-255.
23. Kazazian Jr HH. Mobile elements: Drivers of genome evolution. Science 2004; 303:1626-1632.
24. DeMarco R, Kowaltowski AT, Machado AA et al. Saci-1, -2, and -3 and Perere, four novel retrotransposons with high transcriptional activities from the human parasite Schistosoma mansoni. J Virol 2004; 78(6):2967-78.
25. Casacuberta JM, Santiago N. Plant LTR-retrotransposons and MITEs: Control of Transposition and Impact on the Evolution of Plant genes and genomes. Gene 2003; 311:1-11.
26. Varagona MJ, Purugganan M, Wessler SR. Alternative splicing induced by insertion of retrotransposons into the maize waxy gene. Plant Cell 1992; 4(7):811-20.
27. Goldin SN, Papaioannou VE. Unusual misregulation of RNA splicing caused by insertion of a transposable element into the T (Brachyury) locus. BMC Genomics 2003; 4(1):14.
28. Puig M, Caceres M, Ruiz A. Silencing of a gene adjacent to the breakpoint of a widespread Drosophila inversion by a transposon-induced antisense RNA. Proc Natl Acad Sci USA 2004; 101(24):9013-8.
29. International genome sequencing consortium. Initial sequencing and analysis of the human genome. Nature 2001; 409(6822):860-921.
30. Lippman Z, Gendrel AV, Black M et al. Role of transposable elements in heterochromatin and epigenetic control. Nature 2004; 430(6998):471-6.
31. Evgen'ev MB, Zelentsova, Helena et al. Mobile elements and chromosomal evolution in the virilis group of Drosophila. Proc Natl Acad Sci USA 2000; 97(21):11337-11342.
32. Lim JK, Simmons MJ. Gross chromosome rearrangements mediated by transposable elements in Drosophila melanogaster. Bioessays 1994; 16(4):269-75.
33. Morales ME, Kalinna BH, Heyers O et al. Genomic organization of the Schistosoma mansoni aspartic protease gene, a platyhelminth orthologue of mammalian lysosomal cathepsin D. Gene 2004; 338(1):99-109.
34. Lonnig WE, Saedler H. Chromosome rearrangements and transposable elements. Annu Rev Genet 2002; 36:389-410.
35. Copeland CS, Brindley PJ, Heyers O et al. Boudicca, a retrovirus-like, LTR retrotransposon from the genome of the human blood fluke, Schistosoma mansoni. J Virol 2003; 77:6153-6166.

36. Le Paslier MC, Pierce RJ, Merlin F et al. Construction and characterization of a Schistosoma mansoni bacterial artificial chromosome library. Genomics 2000; 65(2):87-94.
37. Bae YA, Moon SY, Kong Y et al. CsRn1, a novel active retrotransposon in a parasitic trematode, Clonorchis sinensis, discloses a new phylogenetic clade of Ty3/gypsy-like LTR retrotransposons. Mol Biol Evol 2001; 18(8):1474-1483.
38. Copeland CS, Heyers O, Kalinna BH et al. Structural and evolutionary analysis of the transcribed sequence of Boudicca, a Schistosoma mansoni retrotransposon. Gene 2004; 329:103-114.
39. Laha T, Loukas A, Smyth DJ et al. The fugitive LTR retrotransposon from the genome of the human blood fluke, Schistosoma mansoni. International Journal for Parasitology 2004; 34:1365-1375.
40. Michaille JJ, Mathavan S, Gaillard J et al. The complete sequence of mag, a new retrotransposon in Bombyx mori. Nucleic Acids Res 1990; 18(3):674.
41. Laha T, Loukas A, Verity CK et al. Gulliver, a long terminal repeat retrotransposon from the genome of the oriental blood fluke Schistosoma japonicum. Gene 2001; 264(1):59-68.
42. Copeland CS, Mann VH, Morales ME et al. The Sinbad retrotransposon from the genome of the human blood fluke, Schistosoma mansoni, and the distribution of related Pao-like elements. BMC Evol Biol 2005; 5(1):20.
43. Xiong Y, Burke WD, Eickbush TH. Pao, a highly divergent retrotransposable element from Bombyx mori containing long terminal repeats with tandem copies of the putative R region. Nucleic Acids Res 1993; 21:2117-2123.
44. Frame IG, Cutfield JF, Poulter TM. New BEL-like LTR-retrotransposons in Fugu rubripes, Caenorhabditis elegans, and Drosophila melanogaster. Gene 2001; 263:219-230.
45. Fan J, Brindley PJ. Retrotransposable elements in the Schistosoma japonicum genome. In: Isao Tada SK, Moriyasu Tsuji, eds. Ninth International Congress of Parasitology, (August 24-28, 1998). Makuhari Messe. Chiba, Japan: Monduzzi Editore, 1998:821-825.
46. Arkhipova IR, Pyatkov KI, Meselson M et al. Retroelements containing introns in diverse invertebrate taxa. Nat Genet 2003; 33(2):123-4.
47. Foulk BW, Pappas G, Hirai Y et al. Adenylosuccinate lyase of Schistosoma mansoni: Gene structure, mRNA expression, and analysis of the predicted peptide structure of a potential chemotherapeutic target. Int J Parasitol 2002; 32(12):1487-1495.
48. Tammi MT, Arner E, Kindlund E et al. Correcting errors in shotgun sequences. Nucleic Acids Res 2003; 31(15):4663-72.
49. Xiong Y, Eickbush TH. Similarity of reverse transcriptase-like sequences of viruses, transposable elements, and mitochondrial introns. Mol Biol Evol 1988; 5(6):675-90.
50. Boyle JP, Yoshino TP. Gene manipulation in parasitic helminths. Int J Parasitol 2003; 33:1259-1268.
51. O'Brochta D, Sethuraman N, Wilson R et al. Gene vector and transposable element behavior in mosquitoes. J Exp Biol 2003; 206:3823-3834.
52. Feschotte C. Merlin, a new superfamily of DNA transposons identified in diverse animal genomes and related to bacterial IS1016 insertion sequences. Mol Biol Evol 2004.
53. Handler AM. A current perspective on insect gene transformation. Insect Biochem Mol Biol 2001; 31(2):111-28.
54. Garraway LA, Tosi LR, Wang Y et al. Insertional mutagenesis by a modified in vitro Ty1 transposition system. Gene 1997; 198:27-35.
55. Ivics Z, Hackett PB, Plasterk RH et al. Molecular reconstruction of Sleeping Beauty, a Tc1-like transposon from fish, and its transposition in human cells. Cell 1997; 91:501-510.
56. Dewannieux M, Esnault C, Heidmann T. LINE-mediated retrotransposition of marked Alu sequences. Nat Genet 2003; 35:41-48.
57. Ostergaard L, Yanofsky MF. Establishing gene function by mutagenesis in Arabidopsis thaliana. Plant J 2004; 39(5):682-96.
58. Hirochika H. Contribution of the Tos17 retrotransposon to rice functional genomics. Current Opinion in Plant Biology 2001; 4:118-122.
59. Bellen HJ, Levis RW, Liao G et al. The BDGP gene disruption project: Single transposon insertions associated with 40% of Drosophila genes. Genetics 2004; 167:761-81.
60. Williams DC, Boulin T, Ruaud AF et al. Characterization of Mos1 mediated mutagenesis in C. elegans: A method for the rapid identification of mutated genes. Genetics 2005; Epub ahead of print.
61. Amsterdam A, Nissen RM, Sun Z et al. Identification of 315 genes essential for early zebrafish development. Proc Natl Acad Sci USA 2004; 101:12792-7.
62. Isaka K, Nishi H, Sagawa Y et al. Establishment of a new human cell line (EN) with TP53 mutation derived from endometrial carcinoma. Cancer Genet Cytogenet 2003; 141(1):20-5.
63. Stein LD, Bao Z, Blasiar D et al. The genome sequence of Caenorhabditis briggsae: A platform for comparative genomics. PLoS Biol 2003; 1:E45.

64. SanMiguel P, Gaut BS, Tikhonov A et al. The paleontology of intergene retrotransposons of maize. Nat Genet 1998; 20:43-45.
65. Batzer M, Deininger PL. Alu repeats and human genomic diversity. Nat Rev Genet 2002; 3:370-379.
66. Mamedov IZ, Arzumanyan ES, Amosova AL et al. Whole-genome experimental identification of insertion/deletion polymorphisms of interspersed repeats by a new general approach. Nucleic Acids Res 2005; 33(2):e16.
67. Havecker ER, Gao X, Voytas DF. The diversity of LTR retrotransposons. Genome Biol 2004; 5(6):225.
68. Coffin JM, Hughes SH, Varmus HE. Retroviruses. Plainview, NY: Cold Spring Harbor Laboratory Press, 1997.
69. Thompson JD, Gibson TJ, Plewniak F et al. The CLUSTAL_X windows interface: Flexible strategies for multiple sequence alignment aided by quality analysis tools. Nucleic Acids Res 1997; 25:4876-4882.
70. Saitou N, Nei M. The neighbor-joining method: A new method for reconstructing phylogenetic trees. Mol Biol Evol 1987; 4:406-425.

Chapter 7

Non-Long Terminal Repeat Retrotransposons Colonizing Schistosome Genomes

Thewarach Laha, Claudia S. Copeland and Paul J. Brindley*

Abstract

It has become apparent that the schistosome genome is replete with numerous copies of diverse repetitive sequences. Mobile genetic elements constitute much of this repetitive component of the schistosome genome. Of these mobile elements, an expanding number of discrete schistosome retrotransposons—long terminal repeat (LTR) retrotransposons, non-LTR retrotransposons and SINEs—have been identified in recent years. The identity, structure, phylogenetic relationships, and size contributions of the schistosome non-LTR retrotransposons reported to date are reviewed here. (Other chapters in this book focus on LTR retrotransposons of schistosomes [Copeland, Laha, Brindley] and schistosome SINES [Ferbeyre].) These elements include members of the CR1, RTE and R2 clades, including elements termed *SR1*, *SR2*, *SR3*, *pido* and *Perere-3* and *Perere-9* from *Schistosoma mansoni*, *S. japonicum* and *S. haematobium*. An enhanced understanding of these elements will increase our understanding on the evolutionary and phylogenetic relationships of schistosomes and Platyhelminths at large, will facilitate the annotation of the soon-to-be-reported draft genome sequences of *S. mansoni* and *S. japonicum*, and can be expected to have practical implications including the development of transgenesis vectors.

Introduction

Schistosomes are the etiological agents of schistosomiasis, which is considered the most important of the human helminthoses in terms of morbidity and mortality.[1,2] The human schistosomes, blood flukes of the genus *Schistosoma*, possess a comparatively large genome, estimated to be at least 270 megabase pairs (haploid) in size.[3] The nuclear genome is arrayed on eight pairs (2n = 16) of chromosomes including the Z (male) and W (female) sex chromosomes.[4] Transcriptome-based estimates indicate that the genome includes ~14,000 protein encoding genes along with numerous repetitive sequences.[3,5] Much of this repetitive component of the schistosome genome is comprised of mobile genetic elements. Of these mobile elements, an expanding number of discrete retrotransposons - both long terminal repeat (LTR) retrotransposons and non-LTR retrotransposons - have been identified in recent years.[6,7] Members of at least three families of CR1-like, RTE-like and R2-like non-LTR retrotransposons have colonized schistosome genomes, including elements termed *SR1*, *SR2*, *SR3*, *pido* and *perere-3* from *Schistosoma mansoni*, *S. japonicum* and *S. haematobium*. The identity, structure, phylogenetic relationships, and size contributions of the schistosome non-LTR retrotransposons

*Corresponding Author: Paul J. Brindley—Department of Microbiology, Immunology and Tropical Medicine, George Washington University Medical Center, Ross Hall, Room 448, 2300 Eye Street, NW, Washington, DC 20037 USA. Email: mtmpjb@gwumc.edu

Mobile Genetic Elements in Metazoan Parasites, edited by Paul J. Brindley.

are reviewed here. (We have reviewed current information on the LTR retrotransposons from the genomes of schistosomes in a companion chapter of this volume).[7]

Non-LTR Retrotransposons: General Characteristics and Mechanism of Retrotransposition

Non-LTR retrotransposons are about 4 to 6 kb in length (although many copies are usually 5'-truncated and inactive), and encode one or two open reading frames (ORFs). The main ORF (usually the second ORF) encodes a conserved reverse transcriptase (RT) enzyme. In some non-LTR retrotransposons, an apurinic/apyrimidinic endonuclease (APE) coding region is located upstream of the coding sequence of the RT domain. In some others, an RNaseH or a restriction endonuclease-like enzyme domain is located downstream of RT, and some elements also have a protease-like domain.[8] In those non-LTR retrotransposons that encode two ORFs, the first ORF may encode a DNA binding domain.[9] Unlike the LTR retrotransposons, non-LTR retrotransposons lack LTRs and are characterized by A-rich stretches such as poly (A) tails, $(TAA)_n$ or $(TGAAA)_n$ repeats at their 3' ends.[10,11] The absence of LTRs in non-LTR elements clearly indicates different mechanisms of transposition to those used by retroviruses and LTR-retrotranspsosons, although like them non-LTR retrotransposons mobilize via RNA intermediates. Non-LTR elements contain an internal transcription promoter of RNA polymerase II in their sequence.[12]

The mobilization of non-LTR retrotransposons has been reviewed:[13] these elements encode activities necessary for their retrotransposition, and they are termed autonomous even though they probably also require host proteins to complete retrotransposition. In contrast to reverse transcription of LTR retrotransposons and retroviruses, this process takes place on nuclear genomic DNA through target primed reverse transcription. Reverse transcription of non-LTR retrotransposons begins with nicking of the bottom strand of DNA by the EN, leaving a 5'-PO_4 and a 3'-OH. The 3'-OH then serves as a primer with the retrotransposon RNA as template for the RT.[13] Because reverse transcription occurs on the target DNA after cleavage, the process is called target primed reverse transcription, or TPRT. In a variation of TPRT, termed twin priming, inversions are formed, and it is thought that the second strand of DNA is cleaved during reverse transcription of the first strand, and the 3'-OH of the second strand becomes a second primer for reverse transcription internally on retrotransposon RNA. Resolution of this second cDNA produces the inversion.[14] For the L1 elements in the human genome, a new L1 insertion is usually flanked by short direct repeats derived from the target DNA locus upon L1 integration.[15,16] These repeats are called target site duplications (TSDs), and can range from several to several hundred nucleotides in length.[16,17] A schematic diagram of the intracellular life cycle of a generalized non-LTR retrotransposon is presented in Figure 1. Most copies of non-LTR retrotransposons are 5'- truncated and are shorter in length than full length, presumably functional, copies, e.g., 6 kb for human L1. Premature termination of reverse transcription is likely to be responsible.[11] Finally, the nonautonomous SINE-like elements are dependent on the enzymatic machinery of non-LTR retrotransposons for their mobility.[18] Hence, SINES share homologous sequences at their 3' end with non-LTR retrotransposons and are believed to be propagated by the endonuclease and RT of non-LTR retrotransposons.[19]

Phylogeny of Non-LTR Retrotransposons

Eickbush and coworkers have hypothesized that there are at least 12 clades of non-LTR retrotransposons, and that these are ancient elements dating from before the Cambrian era, at least 600 million years ago.[8,9] Malik et al[9] concluded, based on alignments of the RT domain that the CRE clade was the earliest branching clade of non-LTR retrotransposons. CRE includes retrotransposons that are site-specific for mini-exon arrays in trypanosomatid genomes.[20] Other early branching clades include R2 and R4 from arthropods and nematodes, the L1 clade known from algal, slime mold, flowering plant and vertebrate species, and NeSL-1 from *Caenorhabditis elegans*.[8] R2 elements are site-specific for the rRNA genes,

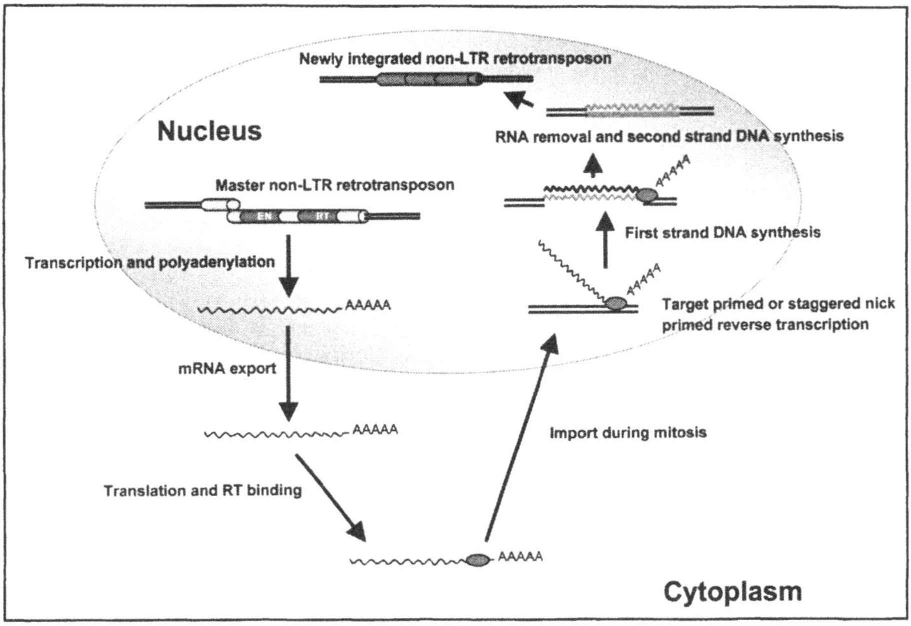

Figure 1. Diagrammatic representation of the replication cycle of a non-LTR retrotransposon. The illustration was assembled from information in references 10-14. (EN, endonuclease; RT, reverse transcriptase).

while members from the R4 clade are site-specific for either rRNA genes or simple repeats. Some representatives from the L1 clade are site-specific (e.g., *Zepp* from the green alga, *Chlorella vulgaris*),[21] whereas others appear to lack site-specificity (e.g., L1 from man). L1 elements apparently integrate at a very large number of sites in the genome because their endonuclease prefers to cleave DNA at a short consensus sequence (5'-TTTT/A-3', where / designates the cleavage site).[13,22,23] The remaining clades of non-LTR retrotransposons are monophyletic within the non-LTR elements. The earliest-diverging clade within these is the RTE clade, with members in nematode, insect, platyhelminth, and vertebrate genomes. Of the other clades, Tad1, LOA and R1 are sister clades, as are CR1 and Jockey. The Tad1 clade has representatives exclusively from fungal genomes, while the LOA, Jockey, and R1 clades may be exclusively present in arthropod genomes. Members of the R1 clade are site-specific either for rRNA genes (e.g., R1 from *Drosophila melanogaster*) or for telomeric repeats (e.g., TRAS from *Bombyx mori*). The CR1 clade is considered at present to have the broadest distribution, with elements identified in insects, nematodes, trematodes (including schistosomes), and vertebrates. The I clade includes representatives from insects, mollusks and trypanosomes.[9]

Schistosome Non-LTR Retrotransposons

Since the first non-LTR retrotransposon was discovered, in the human genome,[24] many members of this class of element have been reported from numerous other organisms.[9,25] A CR1 like non-LTR retrotransposon named *SR1* was the first retrotransposon to be reported from a schistosome genome and in the decade since that report, more than 10 other non-LTR retrotransposons have been characterized from schistosomes, with copy numbers ranging from a few to many thousands (Table 1; Figs. 2,3).[26-30] According to the classification of non-LTR retrotransposable elements of Malik et al[9] these schistosome elements belong to at least three clades of non-LTR retrotransposons including the CR1, RTE-1 and R2 assemblages. Specific

***Table 1.** Estimation by bioinformatics approaches of gene copy number of non-LTR retrotransposons in the genomes of* **Schistosoma mansoni** *and* **S. japonicum**

Host Species	Element Name	Clade	Size (bp)	Estimated Copy Number	% Contribution to Genome Size
S. mansoni	SR1	CR1	~4,500?	200-2000	0.17-1.73
	SR2	RTE	3,913	1,000-10,000	1.45-14.49
	SR3	RTE	3,211	>1,000	1.19-11.89
	Perere	CR1	4,875	250-2500	0.45-4.51
	Perere2	CR1	4,544	480-4800	0.81-8.08
	Perere3	RTE	3,327	2400-24000	2.96-29.57
	Perere4	CR1	5,111	100-1000	0.19-1.89
	Perere5	CR1	5,057	360-3600	0.67-6.74
	Perere6	CR1	4,300	70-700	0.11-1.11
	Perere7	CR1	4,327	400-4000	0.64-6.41
	Perere8	CR1	1,409	150-1500	0.08-0.78
	Perere9	R2	4,394	1-10	0.02-0.16
S. japonicum	Pido	CR1		100	ND
	SjR2	RTE	3,921	1000	ND

ND, not determined.

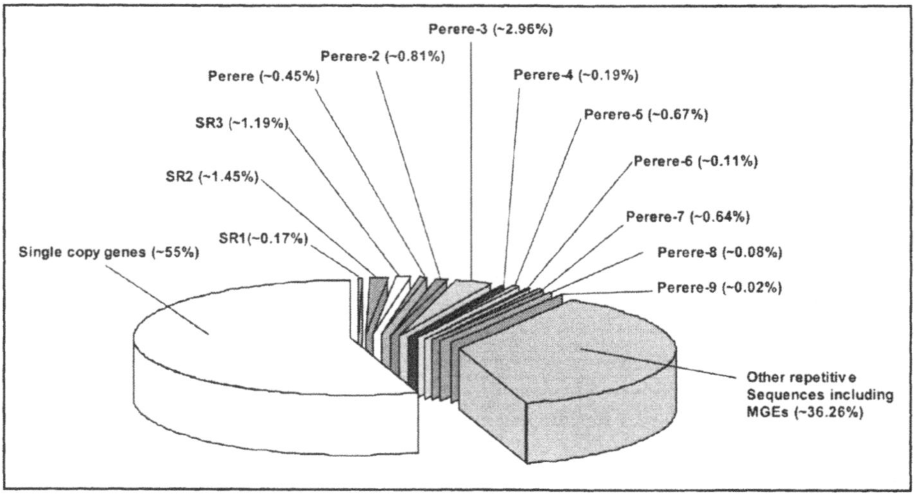

Figure 3. Estimated contribution of non-LTR retrotransposons and other genetic elements in the genome of *Schistosoma mansoni* expressed in pie chart format. Approximate contributions (in terms of simple DNA mass) of the named retrotransposons were predicted deduced from published and gene database sources including (refs. 5-7, 30, 31). See also Table 1.

information on these schistosome non-LTR retrotransposons, including their phylogenetic relationships, is presented below.

CR1-Like Elements

Several CR1 clade LTR retrotransposons have been characterized from schistosomes. These have been named *SR1*, *pido*, *Perere*, *Perere-2*, *Perere-4*, *Perere-5*, *Perere-6* and *Perere-7*.[5,6,26,27,29,30]

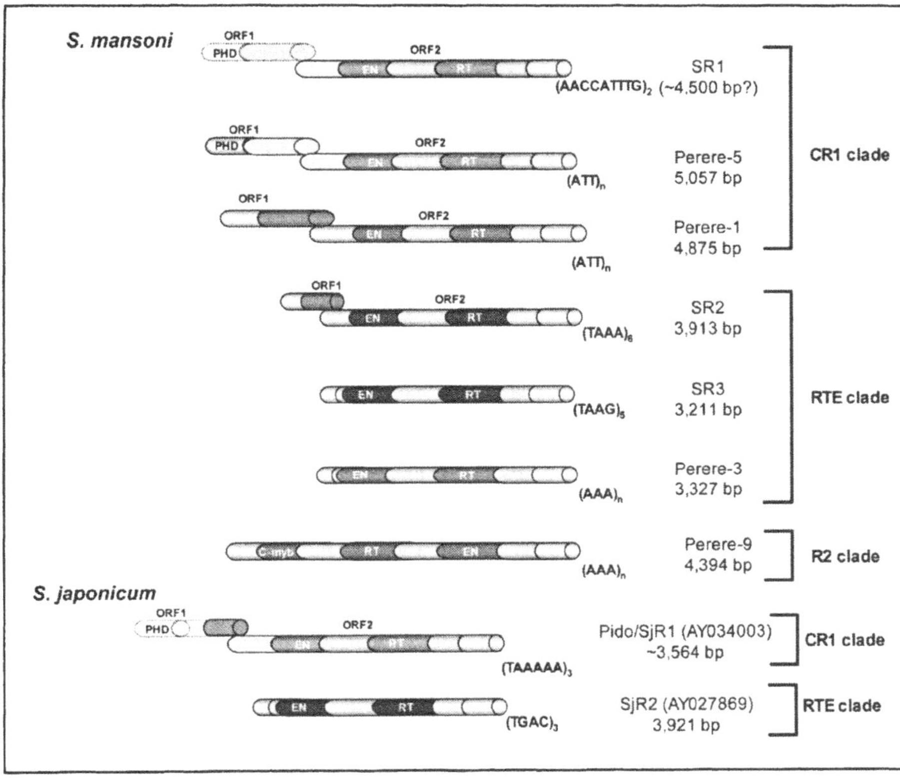

Figure 2. Schematic diagram of representative non-LTR retrotransposons characterized from *Schistosoma* species' genomes including *SR1*, *pido*, *SR2*, *SR3*, *SjR2* and *Perere-9* of the RTE, CR1 and R2 clades. The sequence motifs of the 3'-termini are shown, along with positions of enzymatic domains, EN (endonuclease) and RT (reverse transcriptase) and other functional motifs as appropriate.

SR1 was the first autonomous retrotransposon to be reported from schistosomes.[27] The presence of *SR1* was determined using a PCR based approach to enrich repetitive sequences in the *S. mansoni* genome, and while not apparently of full-length, a consensus sequence for *SR1* was established by screening of partial genomic libraries. Orthologous versions occur in *S. mansoni* (*SR1*) and *S. japonicum* (*SjR1*), with a copy number estimated in the range of 200 to 2,000.[26,28] *SR1* elements possess atypical 3' termini consisting of microsatellite-like repeats, $(AACCATTTG)_2$, which are similar in structure to the imperfect tandem repeat of the 3' termini of *CR1* from chickens and the microsatellite-like 3'-termini of other CR1 clade retrotransposons.[26,27] Subsequent to the report by Drew and Brindley,[26] Kapitonov and Jurka built a 3032-bp consensus sequence of the *SR1* retrotransposon, using genome survey sequences (GSS) from GenBank, that was ~0.7 kb longer than the original sequence.[25] The consensus sequence encoded a 950-aa SR1p protein (positions 36-2885) composed of the APE and RT domains. The extended region encoded APE, but did not permit prediction of any further 5'-extension of the *SR1* consensus sequence, and could not prove or disprove the existence of an additional upstream (first) ORF encoded by this element.

Recently, DeMarco et al,[30] after searching in transcriptome sequences from *S. mansoni*,[3] reconstructed an additional portion of 1,669 bp at the 5' -end of the truncated sequence of Drew and Brindley.[26] The additional segment encoded an endonuclease and an ORF containing a PHD (plant homeodomain) domain, which is a zinc-finger-like motif found in nuclear proteins and which appears to be involved with chromatin-mediated transcriptional regulation.[25] In spite of this attention from several groups, a full length *SR1* element has not yet been described. However, three other SR1-like retrotransposons, *Perere-2*, *Perere-5*, and *Perere-6*, were described by DeMarco et al.[30] These range from 4 kb to 5 kb in length, and all of them contain a PHD domain at the amino-terminus of the first ORF. Thus it can be expected that the ORF1 of SR1 is similar in structure to that of *Perere-2*, *Perere-5*, and *Perere-6*, including an NH_2-terminal PHD domain, and that the full length SR1 retrotransposon is ~4.5 kb in length.

Three additional *S. mansoni* CR1 family retrotransposons, *Perere*, *Perere-4* and *Perere-7*, do not exhibit PHD domains in the deduced proteins coded by their first ORF, so that there exist at least two discrete families of CR1 retrotransposons in the genome of *S. mansoni*, the *SR1* group and the *Perere* group.[30] The sequences of *Perere* (aka *Perere-1*) and *Perere-5* elements terminate at the 3'-end with the motif $(ATT)_n$ (Fig. 2).[30]

A CR1-like element also has been characterized from the genome of the Asian schistosome, *S. japonicum*.[29] This element is termed *pido*. Although the full-length element has yet to be characterized, a consensus sequence of 3,564 bp of the truncated *pido* element has been assembled from several genomic fragments that contained *pido*-hybridizing sequences. The sequence encodes part of the first ORF, the entire second ORF and, at its 3'-terminus, a tandemly repetitive, A-rich motif (Fig. 2). The ORF1 of *pido* encodes a nucleic acid binding protein and ORF2 encodes a retroviral-like polyprotein with APE and RT domains. mRNA encoding the RT of *pido* was detected by reverse transcription-PCR in the egg, miracidium and adult developmental stages of *S. japonicum*, indicating that the RT domain is transcribed and suggesting that *pido* is replicating actively and mobile within the *S. japonicum* genome. At least 1,000 partial copies of *pido* are dispersed throughout the genome of *S. japonicum*. Phylograms based on RT indicate that *pido* and *Perere-4* are orthologues, and if so this would indicate that there are two groupings of *SR1*-like elements in the genome of *S. japonicum* also, as seen in *S. mansoni*—the *SjR1* and the *pido* groups (Fig. 4).

Whereas it has been thought that non-LTR retrotransposons have only been transmitted vertically (and not horizontally),[9] Kapitonov and Jurka[25] have theorized that the CR1 elements are indeed capable of horizontal movement. The first ORF of *CR1* elements contains conserved esterase (ES) or plant homeodomain (PHD) domains, indicating that CR1-like ORF1 proteins are either lipolytic enzymes or are involved in protein-protein interactions related to chromatin remodeling. Sequence conservation of ES suggests that interaction with cellular membranes is an important phase in life cycles of CR1-like elements. Presumably such interaction helps in penetrating host cells. As a consequence, the presence of multiple young CR1 families characterized by >10% intrafamily and 40% interfamily identities may be explained by a relatively frequent horizontal transfer of these CR1-like elements. Further, Kapitonov and Jurka suggest that ES links together non-LTR retrotransposons and single-stranded RNA viruses like influenza C and coronaviruses, which are known to depend on their own ES during horizontal transmission.[25]

RTE-*Like Elements*

Several RTE clade non-LTR retrotransposons have been characterized from schistosomes. These have been named *SR2*, *SjR2*, *SR3* and *Perere-3*.[27,28,30,31] *SR2*, the first schistosome RTE clade element characterized, includes a single, long ORF encoding APE and RT. It is not clear yet if SR2 elements encode an additional ORF upstream of the RT-encoding ORF.[27]

Figure 4, viewed on next page. Phylogram constructed using the neighbor-joining method to compare the relationships among reverse transcriptases of non-LTR retrotransposons belonging to the R2, CR1 and RTE clades of non-LTR retrotransposons from schistosomes and other informative species. Bootstrap values, where 500 or greater from a maximum of 1,000 replicates, are presented at the nodes. See reference 31 for methods used for assembly of the phylogram. Figure legend continued on following page.

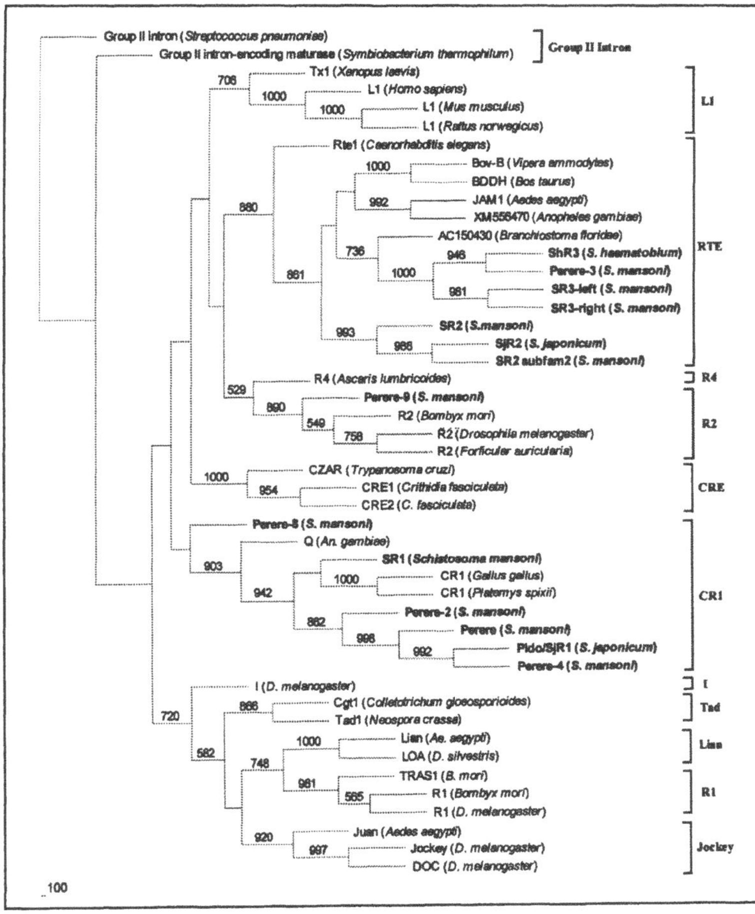

Figure 4, legend continued from previous page. The tree includes family representatives from 11 major clades of non-LTR retrotransposons,[9] as follows: *SR1*(U66331), *SR2* (AF025672), SR2 subfamily B (AF025679), *SR3*-left (DQ008120), SR3-right (DQ008121), *Perere* (BK004067), Perere 2 (BN000793), *Perere-3* (BN000794), Perere-4 (BN000795) and Perere-8 (BN000799) and Perere 9 (BN000800) from *Schistosoma mansoni*, *SjR2* (AY027869) and *pido* (AY034003) from *S. japonicum*, *ShR3* (AY167025) from *S. haematobium*, *RTE1* from *Caenorhabditis elegans* (AF054983), *BDDF* (Z25525) and *BCNT* (AB005652) from *Bos taurus*, JAM1 from *Aedes aegypti* (Z86117), an uncharacterized retrotransposon from *Anopheles gambiae* (XM556470), AC150430 from *Branchiostoma floridae*), Bov-B LINE from *Vipera ammodytes* (AF332697), *R2* from *Bombyx mori* (M16558), *R2* from the earwig, *Forficular auricularia* (AF015819), *R2* from *Drosophila melanogaster* (X51967), *R4* from *Ascaris lumbricoides* (U29445), *CZAR* from *Trypanosoma cruzi* (M62862), *CRE2* from *Crithidia fasciculata* (U19151), *CRE1* from *C. fasciculata* (M33009), *L1* from the rat (U83119), *L1* from the mouse (AF081114), *L1* from the human (U93574), *Tx1* from *Xenopus laevis* (M26915), *CR1* from the turtle *Platymys spixii* (AB005891), *CR1* from the chicken (U88211), *Q* from *Anopheles gambiae* (U03849), *Juan* from *Ae. aegypti* (M95171), *Jockey* (M22874), *Doc* (X17551) and *R1* from *B. mori* (M19755), *R1* from *D. melanogaster* (X51968), TRAS1 from *B. mori* (T18199), *Lian* from *Ae. aegypti* (U87543), *LOA* from *D. silvestris* (X60177), *Tad1* from *Neurospora crassa* (L25662), *CgT1* from the fungal phytopathogen, *Colletotrichum gloeosporioides* (L76169) and *I* (M14954) from *D. melanogaster*. Group II introns and reverse transcriptases from bacteria were used as out groups for the RT tree, specifically group II intron-encoding maturase from *Symbiobacterium thermophilum* (BAD41717; AP006840) and group II intron protein from *Streptococcus pneumoniae* (CAI33690).

Orthologous versions appear to occur in *S. mansoni* (*SR2*) and *S. japonicum* (*SjR2*), with copy numbers estimated in the range of 1,000 to 10,000.[27,28] *SR2* from *S. mansoni* is 3,913 bp, *SjR2* from *S. japonicum* is 3,921 bp. The ORF is bounded by 5'- and 3'-terminal untranslated regions and, at its 3'-terminus, *SjR2* bears a short $(TGAC)_3$ repeat (Figs. 2,4). Active, recombinant RT of *SjR2* has been produced in insect cells.[28]

Laha et al[31] reported the presence of additional RTE-like retrotransposons in a bacterial artificial chromosome (BAC) number 49_J_14 from chromosome 1 of the genome of *S. mansoni* (GenBank AC093105), and therein identified a new non-LTR retrotransposon designated *SR3* (Fig. 5). In addition to the *fugitive* LTR retrotransposon,[32] and a truncated copy of *SR2*, at least two other retrotransposons were readily identifiable in BAC 49_J_14. The first of these was located between nt 346 and 3,552 (i.e., 3,207 bp in length), and the second between nt 97,832 and 101,042 (3,211 bp in length). Comparison of the sequences of these two prospective retrotransposons revealed that they were closely related to one another and appeared to represent discrete copies of a novel family of retrotransposons. Analysis of gene structure and phylogenetic analysis of both the RT and EN domains verified that *SR3* represented a separate family of RTE-like retrotransposons. Remarkably, two full-length copies of *SR3*-like elements were present in BAC 49_J_14, and one of 3,211 bp in length appeared to be intact, indicating *SR3* is an active non-LTR retrotransposon. Southern hybridization and bioinformatics analyses indicated the presence of numerous copies (probably >1,000) of *SR3* interspersed throughout the genome of *S. mansoni*. Bioinformatics analyses also revealed *SR3* to be transcribed in both larval and adult developmental stages of *S. mansoni* and to be also present in the genomes of the other major schistosome parasites of humans, *Schistosoma haematobium* and *S. japonicum*. Further, Laha et al[31] investigated the nature of integration sites or target sequences of *SR3* within the schistosome genome by employing five kb of nucleotide sequences flanking the 5'- and 3'- UTRs of both *SR3-left* and *SR3-right* to query the GenBank nonredundant nucleotide and protein databases, and the GSS and EST entries for *S. mansoni*. While the searches revealed no significant matches to any sequences encoding genes of *Schistosoma* species, they did reveal that *SR3* elements appear to target AT-rich sites, as indicated in Figure 6, a similar preference to *L1* retrotransposons within the human genome.[16,22] More specifically, the average AT content of the integration sites of the 21 copies of *SR3* shown in Figure 6 was 68 % AT.[31] Whereas target site specificity does not appear to be stringent for *SR3*, it can be expected to reflect the recognition sequence of the *SR3* endonuclease. For example, L1 elements apparently integrate at numerous sites in the genome because the endonuclease of L1 preferably cleaves DNA at the short consensus sequence, 5'-TTTT/A-3', where / designates the cleavage site.[22,23]

To propagate, non-LTR retrotransposons employ their EN and RT enzymes respectively to nick a genomic target site and reverse transcribe the retrotransposon, integrating the element into a new genomic locus (Fig. 1). As noted above, this process is termed target-site-primed reverse transcription. For the L1 elements in the human genome, a new L1 insertion is usually flanked by short direct repeats derived from the target DNA locus upon L1 integration.[13] These repeats are called target site duplications (TSDs), and can range from several to several hundred nucleotides in length.[16,17] *SR3-left* and *SR3-right* are flanked by TSDs of 10 and 12 bp, respectively; TAGTGGCTAATCT for *SR3-right* and CGCTCTTAAA for *SR3-left* (Fig. 6). The presence of TSDs provided indication, along with their intact structure, of recent activity of these two copies of *SR3* localized in BAC 49_J_14.[17,31] Apparently unlike *SR3*, and certainly unlike L1, some other clades of non-LTR retrotransposons exhibit extreme target site specificity, the well-known examples being the *R2* and *R4* elements which are found in the ribosomal RNA genes of insects (e.g., *Bombyx mori, Drosophila melanogaster*), nematodes (e.g., *Ascaris lumbricoides*), and in *S. mansoni* (see *Perere-9*, below), or in simple repeats (e.g., the *Dong* element from *B. mori*).[9] Interestingly, an *SR3* of *S. haematobium* occurs in the introns of the AChE gene, and in addition, other RTE elements have been reported from gene-rich sites of the schistosome genome.[31] The degenerate copy of a non-LTR retrotransposon, *SR2* in BAC clone BAC 49_J_14 (Fig. 5) has integrated into intron 1 of the Zn-Cu superoxide dismutase

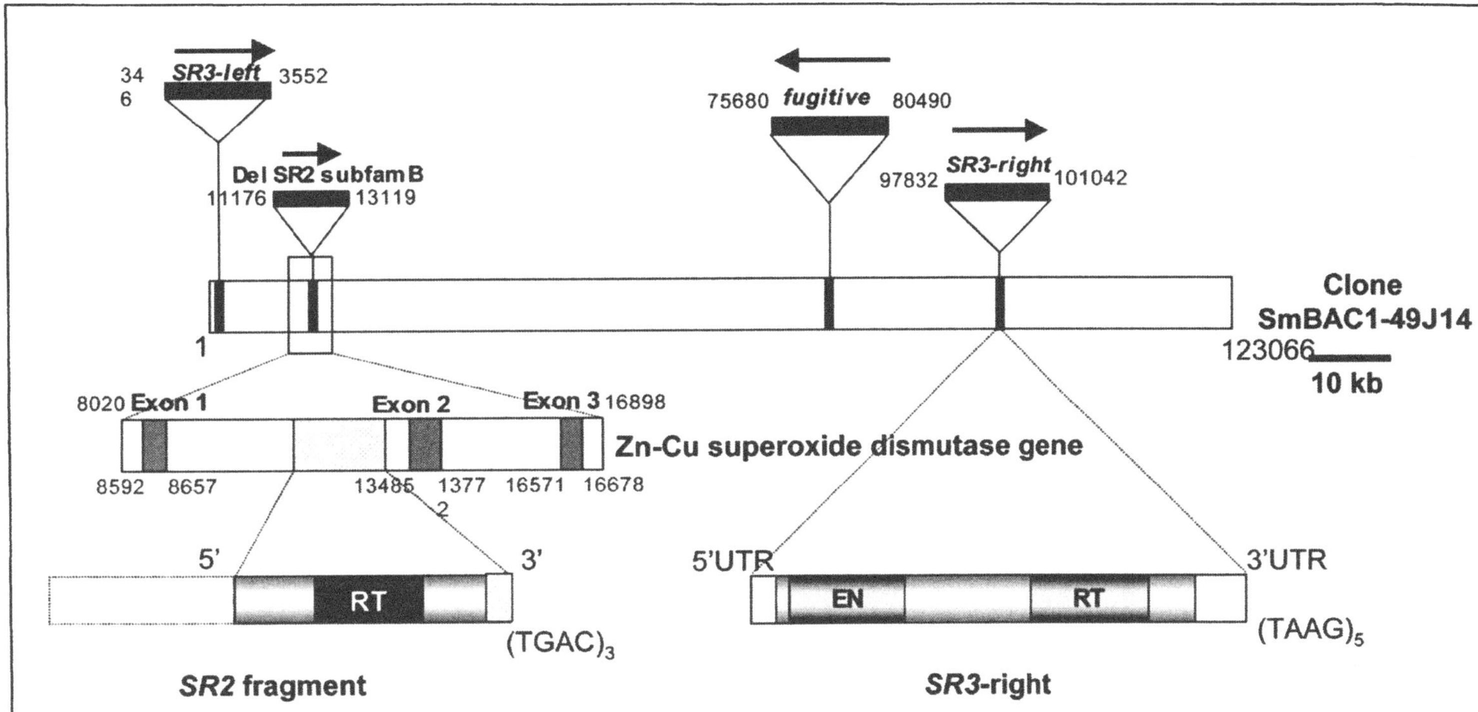

Figure 5. Schematic diagram of the location, size and structure of copies of the *SR3* non-LTR retrotransposon in bacterial artificial chromosome number 49_J_14 from chromosome 1 of the genome of *Schistosoma mansoni*. The location of copies of the *SR2* and *fugitive* retrotransposons is also presented. The arrows indicate the direction of transcription of the mobile elements. The numbers on each bar indicate the nucleotide position of the elements within the bacterial artificial chromosome. A degenerate copy of the *SR2* element was evident within intron 1 of Zn-Cu superoxide dismutase (SOD) gene, in particular within intron 1 of the gene: three exons of the SOD are indicated in red, with the position and structure of the degenerate copy of *SR2* indicated within intron 1. On the bottom right, a schematic of the length and domain structure of the *SR3-right* copy of the *SR3* retrotransposon is presented. RT, reverse transcriptase; EN, endonuclease. (Adapted from Laha et al.[31])

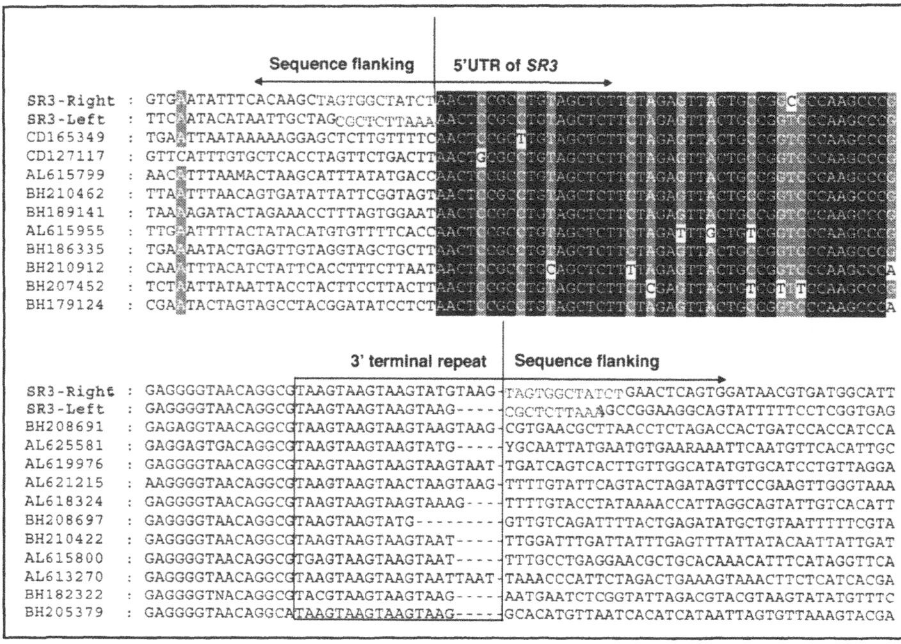

Figure 6. Multiple sequence alignment of the nucleotide sequences flanking the insertion sites of copies of the *SR3* non-LTR retrotransposon within the genome of *Schistosoma mansoni*. Sequences flanking the 5'UTR of *SR3-left* and *SR3-right* are aligned in the top panel, while those flanking the 3'-terminus of *SR3-left* and *SR3-right* are presented in the bottom panel. Target site duplications are evident at the sites of *SR3-left* and *SR3-right* integration, and are indicated with blue colored font. Conservation of residues is indicated by the shading of boxes. Target sequences were identified among entries in the GSS database of *S. mansoni* sequences at GenBank or the Sanger Institute (http://www.ncbi.nlm.nih.gov/dbGSS/index.html; http://www.genedb.org/genedb/smansoni/blast.jsp; http://www.sanger.ac.uk/cgi-bin/blast/submitblast/s_mansoni). (Adapted from Laha et al.[31])

gene of *S. mansoni*. *SR2* from schistosomes has been recorded from several other target genes including 28 kDa glutathione *S* transferase, cathepsin D and the UTR of heat shock protein.[27,28,33] Thus, although *SR3* and other RTE clade retrotransposons do not exhibit tight target site specificity, they seem to prefer to integrate into AT-rich sites and, in addition, are frequently found in introns and other-noncoding areas of protein encoding gene loci. Finally, a closely related RTE like element, termed *Perere-3*, has been described to have a highly variable set of extreme 3'terminal sequences,[30] although phylogenetic analysis of the RT sequence of *Perere-3* demonstrated that it is an *SR3*-like retrotransposon in sequence and structure (Figs. 2, 4).

R2-Like Elements

R2 is an early branching clade of non-LTR retrotransposons.[9] R2 elements are target site specific retrotransposons known primarily from the genomes of arthropods and nematodes. R2 elements have recently been described from *S. mansoni*.[30,34] In particular, after searches of public genomic sequences, Kojima and Fujiwara identified R2-like sequences, and reconstructed two R2 elements, designated R2Sm-A and R2Sm-B. R2Sm-A shared structural features known from other R2 elements including a long, single open reading frame that included at its NH_2-terminal end a pair of CCHH zinc finger motifs, a c-myb-like DNA-binding motif, RT, and at its COOH-end, a CCHC zinc finger domain and finally a restriction enzyme-like endonuclease.

Intriguingly, an R2Sm-A like element was identified in a ribosomal gene from a nonhuman schistosome,[35] where it was found to have inserted into a 28S ribosomal gene apparently at the consensus 5'-AAGG / TAGC-3' known for other R2 elements within ribosomal genes of other host genomes. Based on its structure and sequence, Kojima and Fujiwara proposed that R2Sm-A from schistosomes formed a discrete family of R2 elements that they termed R2-C.[34]

Second, DeMarco et al[30] obtained by PCR a genomic clone of a novel retrotransposon that they termed *Perere-9*. Although *Perere-9* appeared to be an inactive copy based on presence of mutations,[30] phylogenetic analysis confirmed that *Perere-9* was a R2 family non-LTR retrotransposon. Alignment of the R2 protein coded by the reconstructed copy of *Perere-9* along with several R2 proteins from retrotransposons of other species showed that it exhibits the conserved NH_2-terminal CCHH and c-myb domains,[36] and a characteristic COOH-terminal endonuclease domain with a CCHC/PD..D motif known from the R2, R4, and CRE clades of non-LTR retrotransposons.[37] Like Kojima and Fujiwara,[34] DeMarco et al[30] also noted the presence of copies of *Perere-9* -like sequences in ribosomal genes of *S. mansoni* (e.g., GenBank CF497869), indicating a target specificity for *Perere-9* that is similar to that known for the *Dong* retrotransposon of *Bombyx mori*.[36] Finally, there appear to be only a few copies of *Perere-9* in the genome of *S. mansoni*, which is in contrast to the RTE-like schistosome retrotransposons that are present at high copy number (Table 1; Fig. 3).

Non-LTR Retrotransposon Processed Pseudogenes?

In addition to the role of RT in the retrotransposition of the cognate non-LTR retrotransposon, the activity of RT encoded by non-LTR retrotransposons might facilitate retrotransposition of other genes as known from other organisms.[38] This phenomenon could affect heterologous genes including SINEs and processed pseudogenes. While there is as of yet no evidence in schistosomes to definitively implicate non-LTR retrotransposons in generating processed pseudogenes, this phenomenon is known with human LINE retrotransposons[38] where processed pseudogenes occur with loss of intron and promoter sequences, acquisition of a polyA at their 3' end and target site duplication. In addition to non-LTR retrotransposon generated, processed pseudogenes, the RT activities in trans of schistosome non-LTR retrotransposon can be expected to be responsible for the movement of *Sm-α* and other schistosome SINEs.[18,39-41]

Acknowledgements

We acknowledge financial support to TL from the UNICEP/UNDP/World Bank/WHO Special Programme for Research and Training in Tropical Diseases (TDR) (ID A20723). PJB is a recipient of a Burroughs Wellcome Fund scholar award in Molecular Parasitology.

References

1. King CH, Dickman K, Tisch DJ. Reassessment of the cost of chronic helmintic infection: A meta-analysis of disability-related outcomes in endemic schistosomiasis. Lancet 2005; 365:1561-1569.
2. Hotez PJ, Molyneux DH, Fenwick A et al. Incorporating a rapid-impact package for neglected tropical diseases with programs for HIV/AIDS, tuberculosis, and malaria. PloS Medicine 2006; 3:e102.
3. Verjovski-Almeida S, DeMarco R, Martins EA et al. Transcriptome analysis of the acoelomate human parasite Schistosoma mansoni. Nat Genet 2003; 35:148-157.
4. Hirai H, Taguchi T, Saitoh Y et al. Chromosomal differentiation of the Schistosoma japonicum complex. Int J Parasitol 2000; 30:441-452.
5. DeMarco R, Kowaltowski AT, Machado AA et al. Saci-1, -2, and -3 and Perere, four novel retrotransposons with high transcriptional activities from the human parasite Schistosoma mansoni. J Virol 2004; 78:2967-78.
6. Brindley PJ, Laha T, McManus DP et al. Mobile genetic elements colonizing the genomes of metazoan parasites. Trends Parasitol 2003; 19:79-87.
7. Copeland CS, Laha T, Brindley PJ. Schistosome long terminal repeat retrotransposons. In: Brindley PJ, ed. Mobile Genetic Elements in Metazoan Parasites. Austin: Landes Bioscience, 2009: Chapter 6, this volume.
8. Malik HS, Eickbush TH. NeSL-1, an ancient lineage of site-specific non-LTR retrotransposons from Caenorhabditis elegans. Genetics 2000; 154:193-203.

9. Malik HS, Burke WD, Eickbush TH. The age and evolution of non-LTR retrotransposable elements. Mol Biol Evol 1999; 16:793-805.
10. Eickbush TH. Transposing without ends: The non-LTR retrotransposons. New Biol 1992; 4:430-440.
11. Finnegan DJ. Transposable elements: How non-LTR retrotransposons do it. Curr Biol 1997; 7:R245-R248.
12. Kazazian Jr HH, Moran JV. The impact of L1 retrotransposons on the human genome. Nat Genet 1998; 19:19-24.
13. Kazazian Jr HH. Mobile elements: Drivers of genome evolution. Science 2004; 303:1626-1632.
14. Ostertag EM, Kazazian Jr HH. Twin priming: A proposed mechanism for the creation of inversions in L1 retrotransposition. Genome Res 2001; 11:2059-2065.
15. Fanning TG, Singer MF. LINE-1: A mammalian transposable element. Biochim Biophys Acta 1987; 910:203-212.
16. Szak TS, Pickeral OK, Makalowski W et al. Molecular archeology of L1 insertions in the human genome. Genome Biol 2002; 3:10.
17. Moran JV, Holmes SE, Naas TP et al. High frequency retrotransposition in cultured mammalian cells. Cell 1996; 87:917-927.
18. Ferbeyre G. Transcripts from Schistosoma nonautonomous retrotransposons possess catalytic activity. In: Brindley PJ, ed. Mobile Genetic Elements in Metazoan Parasites. Austin: Landes Bioscience, 2009: Chapter 1, this volume.
19. Deininger PL, Batzer MA. Mammalian retroelements. Genome Res 2002; 12:1455-1465.
20. Teng SC, Wang SX, Gabriel A. A new non-LTR retrotransposon provides evidence for multiple distinct site-specific elements in Crithidia fasciculata miniexon arrays. Nucl Acids Res 1995; 23:2929-2936.
21. Higashiyama T, Noutoshi Y, Fujie M et al. Zepp, a LINE-like retrotransposon accumulated in the Chlorella telomeric region. EMBO J 1997; 16:3715-3723.
22. Feng Q, Moran JV, Kazazian Jr HH et al. Human L1 retrotransposon encodes a conserved endonuclease required for retrotransposition. Cell 1996; 87:905-916.
23. Jurka J. Sequence patterns indicate an enzymatic involvement in integration of mammalian retroposons. Proc Natl Acad Sci USA 1997; 94:1872-1827.
24. Hutchison CA, Hardies SC, Loeb DD et al. LINEs and related retroposons: Long interspersed repeated sequences in the eukaryotic genome. In: Berg DH, Howe MM, eds. Mobile DNA. American Society of Microbiology. Washington, DC: 1989:593-617.
25. Kapitonov VV, Jurka J. The esterase and PHD domains in CR1-Like non-LTR retrotransposons. Mol Biol Evol 2003; 20:38-46.
26. Drew AC, Brindley PJ. A retrotransposon of the nonlong terminal repeat class from the human blood fluke Schistosoma mansoni. Similarities with the chicken repeat 1-like elements from vertebrates. Mol Biol Evol 1997; 14:602-610.
27. Drew AC, Minchella DJ, King LT et al. SR2, nonlong terminal repeat retrotransposons of the RTE-1 lineage, from the human blood fluke Schistosoma mansoni. Mol Biol Evol 2005; 16:1256-1269.
28. Laha T, Brindley PJ, Smout MJ et al. Reverse transcriptase activity and UTR sharing of a new RTE-like, non-LTR retrotransposon from the human blood fluke, Schistosoma japonicum. Int J Parasitol 2002; 32:1163-1174.
29. Laha T, Brindley PJ, Verity CK et al. pido, a nonlong terminal repeat retrotransposon of the chicken repeat 1 family from the genome of the Oriental blood fluke, Schistosoma japonicum. Gene 2002; 284:149-159.
30. DeMarco R, Machado AA, Bisson-Filho AW et al. Identification of 18 new transcribed retrotransposons in Schistosoma mansoni. Biochem Biophys Res Commun 2005; 333:230-240.
31. Laha T, Kewgrai N, Loukas A et al. Characterization of SR3 reveals abundance of non-LTR retrotransposons of the RTE clade in the genome of the human blood fluke, Schistosoma mansoni. BMC Genomics 2005; 6:154.
32. Laha T, Loukas A, Smyth DJ et al. The fugitive LTR retrotransposon from the genome of the human blood fluke, Schistosoma mansoni. Int J Parasitol 2004; 34:1365-1375, (Corrigendum - Int J Parasitol. 2005; 35:461).
33. Morales ME, Kalinna BH, Heyers O et al. Genomic organization of the Schistosoma mansoni aspartic protease gene, a platyhelminth orthologue of mammalian lysosomal cathepsin D. Gene 2004; 338:99-109.
34. Kojima KK, Fujiwara H. Long-term inheritance of the 28S rRNA specific retrotransposon R2. Mol Biol Evol 2005;22:2157-2165.
35. Morgan JA, DeJong RJ, Kazibwe F et al. A newly-identified lineage of Schistosoma. Int J Parasitol 2003; 33:977-985.

36. Burke WD, Malik HS, Jones JP et al. The domain structure and retrotransposition mechanism of R2 elements are conserved throughout arthropods. Mol Biol Evol 1999; 16:502-511.
37. Yang J, Malik HS, Eickbush TH. Identification of the endonuclease domain encoded by R2 and other site-specific, nonlong terminal repeat retrotransposable elements. Proc Natl Acad Sci USA 1999; 96:7847-7852.
38. Esnault C, Maestre J, Heidmann T. Human LINE retrotransposon generate processed pseudogenes. Nat Genet 2000; 24:363-367.
39. Spotila LD, Hirai H, Rekosh DM et al. A retroposon like short repetitive DNA element in the genome of the human blood fluke, Schistosoma mansoni. Chromosoma 1989; 97:421-428.
40. Laha T, McManus DP, Loukas A et al. Sjα elements, short interspersed element-like retroposons bearing a hammerhead ribozyme motif from the genome of the Oriental blood fluke Schistosoma japonicum. Biochim Biophys Acta 2000; 1492:477-482.
41. Malik HS, Eickbush TH. The RTE class of non-LTR retrotransposons is widely distributed in animals and is the origin of many SINEs. Mol Biol Evol 1998; 15:1123-34.

CHAPTER 8

Schistosome DNA Transposons

Ricardo DeMarco* and Sergio Verjovski-Almeida

Abstract

Schistosomes are digenetic blood flukes and the causative agents of schistosomiasis in humans. Large transcriptome and genome sequence databases have recently become publicly available for *Schistosoma mansoni* and *S. japonicum* and have provided a comprehensive repository for the discovery of novel repetitive elements. In fact, two DNA transposons, namely Merlin_Sm1 and SmTRC1, have been recently described in *S. mansoni* and are the first classical transposons to be described in platyhelminths. This chapter reviews the characterization of each of these elements in relation to structure, transcription and putative encoded proteins, and provides a critical analysis on the potential use of these two elements as tools for transgenesis and mutagenesis of schistosomes.

Introduction

Transposable Elements and DNA Transposons

Transposable elements are discrete segments of DNA that are characterized by their mobility within genomes. They display a replication capacity permitting the colonization of genomes and, in some cases, have been implicated in major variations in genome size and organization in a short evolutionary period.[1,2] They can be deleterious to the host by inserting in genes or gene regulatory regions or by inducing chromosomal rearrangements by recombination being initially regarded as selfish elements.[3] In contrast, there are several examples where these elements show an altruistic behavior, such as contributing to gene evolution,[4-6] DNA repair[7] and, in the one of the most extreme examples, being responsible for telomeric maintenance in the genus *Drosophila*.[8] Transposable elements can be classified in two broad groups, namely retroelements (Class I) and DNA transposable elements (Class II). Unlike retroelements, DNA transposons are not restricted to eukaryotes, being also widespread in prokaryote organisms.

Class II DNA transposable elements use a transposase for a multi-step cut and paste transposition reaction directly from DNA. The DNA transposon element is usually delimited by the Terminal Inverted Repeats (TIRs), which provide the site for binding of the transposase that will catalyze the cleavage and transfer reaction of the element. Autonomous DNA transposable elements encode at least one protein, the transposase. Integration of the element comprises the nucleophilic attack of the 3'OH ends of the free transposon to each strand of target DNA in positions few bases apart, followed by repair by host machinery of the single strand regions flanking the transferred element.[9] This process results in target-site duplication (TSD), which is evident as a direct repeat flanking each of the TIRs of the newly integrated element. As the size of this direct repeat is linked to the mechanism of insertion of the transposon by the

*Corresponding Author: Ricardo DeMarco—Departamento de Física e Informática, Instituto de Física de São Carlos, Universidade de São Paulo, 13566-590 São Carlos, SP, Brazil.
Email: rdemarco@ifsc.usp.br

Mobile Genetic Elements in Metazoan Parasites, edited by Paul J. Brindley.

transposase, each family of transposons usually displays a distinctive TSD length. Analyzing the amino acid sequence similarity of the encoded transposases, eukaryote DNA transposons can be divided into nine superfamilies: Mariner-TC1; hAT; P; Mutator; CACTA; PIF/Harbinger; Transib; *piggyBac* and Merlin.[10-12]

Schistosomes and Transposable Elements

The schistosomes are digenetic blood flukes, and the causative agents of schistosomiasis in humans. The disease is endemic in 74 developing countries infecting about 200 million individuals and is estimated that an additional 500-600 million are at risk.[13] The main schistosome species currently infecting human beings are *S. mansoni* which causes intestinal and hepatic schistosomiasis in Africa, the Arabian peninsula, and South America; *S. haematobium* causing urinary schistosomiasis in Africa and the Arabian peninsula; and *S. haematobium*, causing intestinal and hepatosplenic schistosomiasis in China, the Philippines, and Indonesia.[14]

Schistosomes display a complex life cycle, infecting an intermediate (snail) host and a permanent (vertebrate) host. The adult worms living in the vertebrate veins mate and lay eggs that pass the intestinal (*S. mansoni* and *S. japonicum*) or the bladder (*S. japonicum*) wall and are released to the medium in the feces or urine, respectively. The eggs hatch in fresh water, releasing a free-swimming miracidium, which penetrates into the snail host. In this intermediate host, the *Schistosoma* will differentiate into a sporocyst and by means of asexual reproduction will produce thousands of cercariae. These will be released into the environment and penetrate into the vertebrate host skin. After penetration, they differentiate into schistosomula and travel through the skin to a venule, from where they will direct to the lungs and undergo further developmental changes. Finally after 8 to 10 days the parasites will migrate to the liver where they will mature into adult worms.

S. mansoni genome was initially estimated to be 270 Mbp in size,[15] but recent data from the genome sequencing project suggests a size of approximately 300 Mbp.[16] Approximately 55% of the genome content is expected to be composed of mobile elements or other repetitive sequences.[17] In fact, the high repetitive content of *S. mansoni* genome still represents a considerable obstacle to assemble the genome sequence at chromosome level.[16] The recent large scale sequencing of *S. mansoni*[18] and *S. japonicum*[19] transcriptomes, and the release of *S. mansoni* genomic sequence preliminary data assembly[16] provided a large repository for the discovery of novel repetitive elements.

Several families of retroelements have been described in *Schistosoma* and a large portion of their genome is derived from such elements;[17,20-23] the retroelements are addressed in two companion chapters in this book. In contrast, only two DNA transposons have been described so far in schistosomes, both of them in *S. mansoni*: the Merlin_Sm1[12] and the SmTRC1[24] (Fig. 1), and they are the subject of the present chapter. The estimated contribution of these two elements in the composition the schistosome genome is relatively small compared with retroelements. Merlin_Sm1 was estimated by in silico searches to have approximately 500 copies[12] and SmTRC1 was estimated by Southern blot experiments to display between 30 to 300 copies,[24] representing in total less than 1% of the *S. mansoni* genome. This scenario is similar to the described for several organisms, such as *Homo sapiens, Mus musculus* and *Anopheles gambiae,* which also display a much higher proportion of retroelements in their genomes when compared with DNA transposable elements.[25]

Schistosomes DNA Transposons

Merlin_Sm1

The Merlin superfamily of DNA transposons has been recently characterized, being detected in a wide range of animals. They display TIRs from 24 to 462 bp long and generate an 8 or 9 bp TSD. A protein of approximately 300 amino acids is encoded by these elements and it displays similarities with a transposase from Bacterial IS1016 Insertion Sequences, pointing to a relationship between those two families.[12]

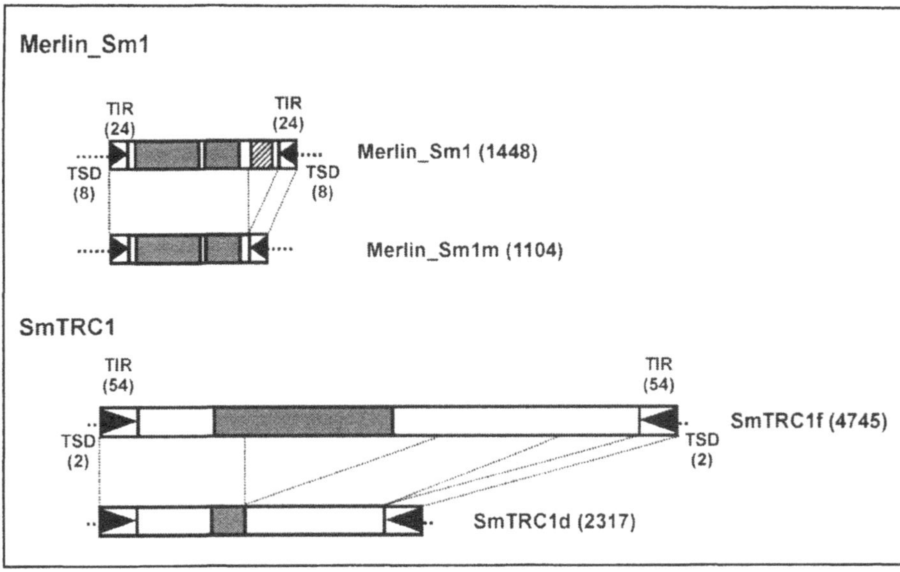

Figure 1. Schematic representation of *S. mansoni* DNA transposons. Black triangles represent the Terminal Inverted Repeats (TIRs). Gray boxes represent the predicted open reading frame (ORF) for each element (for Merlin_Sm1 the ORF is interrupted by a predicted short 32bp intron, represented by a thin white line within the gray ORF). Thick dotted lines represent the direct repeat generated by target-site duplication (TSD). Thin dotted lines connect orthologous regions between truncated and full length elements. The hatched box represents the region within Merlin_Sm1 with similarity to Perere-10 Penelope-like element (PLE) transposon, detected in the present work. Numbers in parentheses indicate the length in base pairs of each element or feature.

Merlin_Sm1 was the first DNA transposon described in flatworms. It was detected by Feschotte[12] using an in silico approach. BLAST searches using a merlin tranposon from the nematode *Caenorhabiditis briggsae* as query were performed against the *S. mansoni* whole genome preliminary sequence database at the Sanger Institute, which detected a canonical element named Merlin_Sm1 with 1448 bp and 24 bp TIRs in its extremities. Using the assembled genome information at least 500 copies, with greater than 80% similarity, have been detected[12] and most of the copies displayed an internal deletion in relation to the canonical copy (Fig. 1). Merlin-like sequences were also detected in the *S. japonicum* EST database at GenBank, pointing to the existence of a Merlin transposon in this species.

The Merlin_Sm1 is predicted to generate a transcript that is composed of two exons separated by an intron of 32 bp (Fig. 2, gray bar). This transcript would encode a 294 aa putative transposase, with 34% identity and 56% similarity with Merlin_Cb1p. However, it is interesting to note that a search in the 124,000 reads generated by the *S. mansoni* transcriptome project only detected the presence of spliceless transcripts[12] (Fig. 2, black bar); all 3 *S. mansoni* ESTs that mapped in the intron region represent unspliced transcripts whereas no EST representing a spliced transcript was detected. In contrast, 4 out of 5 *S. japonicum* ESTs from GenBank mapping in this area represent transcripts displaying a spliced message (Fig. 2, Merlin_Sm1 light gray bar).

The intronic sequence is only 68% identical between the two species; however the canonical splicing sites GT-AG can be detected in 2 of the *S. mansoni* ESTs at the positions where the extremities of the intron would be located. Retention of the intron leads to a frameshift and encodes a putative truncated protein.[12] Therefore, it should be expected that a yet undetected spliced version of this transcript is generated in *S. mansoni*, encoding the complete non-truncated transposase. Additionally, we note that all unspliced messages (including the one from *S. japonicum*) were derived from adult worms, and all spliced messages were obtained from egg RNA. These data suggest that splicing of this transcript may be differentially regulated throughout the schistosome life cycle and that the full transposase may only be produced in some stages.

It has been suggested that the truncated Merlin transposase might exert some inhibitory effect on transposition,[12] in a similar fashion as seen for P element, which transposes at high rates in the germline but not in somatic tissue. The germline specificity of P element transposition is due to a germline-specific splice that joins two open reading frames and produces an integral transposase, while the somatic tissue only produces a transposition repressive truncated transposase.[26] A more comprehensive study is warranted to investigate the differential expression of the two splice variants of this transcript throughout the schistosome life cycle and to determine if the Merlin_Sm1 truncated transposase has in fact some repressor function analogous to that seen in the P element.

Analyzing the 3' portion of a 1,448 bp canonical copy of Merlin_Sm1 obtained from the v3.0 assembly of the *S. mansoni* genome,[16] we now note that the portion that is unique to the canonical copy, and is absent from the deletion derivatives (Fig. 1, hatched box), has a high degree of identity (95% over 284 nucleotides) with Perere-10, a recently described Penelope-like transposon.[21] It has been described that in the canonical copies a direct repeat CATTTAAG flanks this deleted internal portion, and it was hypothesized that such repeats may have caused a slippage during replication or repair that caused the deletion.[12] Considering the identity of the deleted portion, it is now possible to propose an alternative hypothesis, in which the longer copy was generated by the insertion of a Perere-10 like element into the CATTTAAG site of a shorter 1104 bp Merlin_Sm1 element, which would have caused a target-site duplication (TSD) flanking the transposon, thus explaining the origin of such direct repeats in the longer copy. The length of this putative TSD (8 bp) is similar to the 7 bp TSD that has been determined at the insertion site of *Drosophila* Penelope transposon.[27]

SmTRC1

The CACTA superfamily of DNA transposons (also called En/Spm) was named after the core sequence CACTA displayed in their TIRs, and the presence of elements from this family has been described in several plants. The best known member of this family is the maize Suppressor-mutator Spm transposon, which was first described by Barbara McClintock.[28] Further molecular studies determined the En/Spm sequence[29] and showed that most of the transposon length is occupied by a single transcription unit, which can be alternatively spliced to generate four distinct transcripts (*tnpA* to *tnpD*) encoding different proteins.[30] Recently, novel families of transposons distantly related to the CACTA elements of plants have been described in several metazoan and fungal species, and termed TRC for Transposons Related to CACTA.[24]

Analyzing the gene structure for a certain transcript mapping to the *S. mansoni* genome, DeMarco et al[24] noticed a 4.5 kbp repetitive element, which was named SmTRC1, present in one of the transcript introns. An inverted repeat of 54 bp delimitated this element that displays a 1683 bp ORF (Fig. 1). Several SmTRC1 elements detected in the *S. mansoni* genome display a 2 bp direct repeat flanking the TIRs, indicating a target-site duplication TSD event during the insertion of these elements.[24] PCR amplification of *S. mansoni* DNA using primers designed at the extremities of SmTRC1 followed by cloning and sequencing[24] confirmed the existence of copies of SmTRC1 with integral and truncated structures; the general structure of these elements is shown in Figure 1 (SmTRC1f and SmTRC1d, respectively).

In silico analysis of the protein encoded by SmTRC1 ORF showed a limited similarity to transposases of plant CACTA elements. Additional in silico searches with SmTRC1 ORF product revealed several additional TRC elements in animals and fungi.[24] Sequence alignment using know CACTA transposons from plants and the novel TRC elements shows

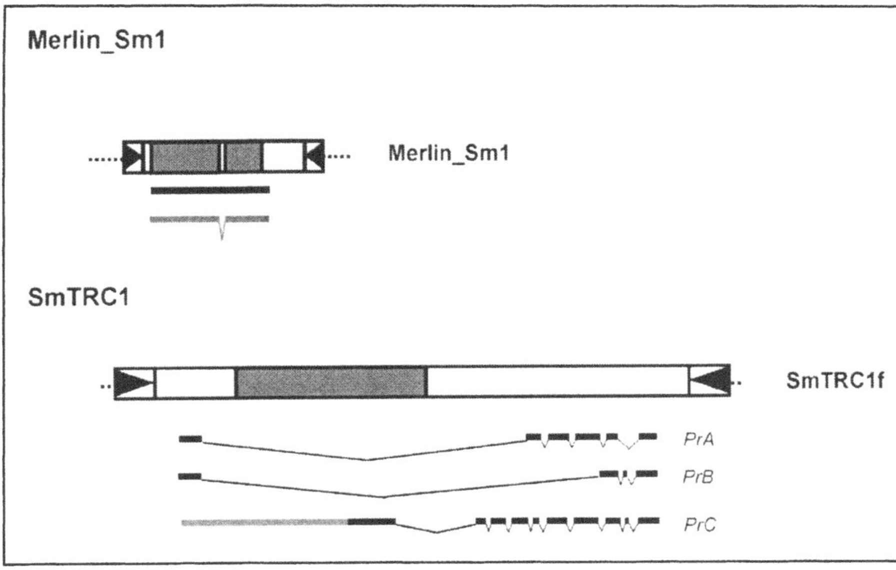

Figure 2. Schematic representation of *S. mansoni* DNA transposon transcripts. Black triangles represent the Terminal Inverted Repeats (TIRs). Gray boxes represent the predicted ORF for each element (for Merlin_Sm1 the ORF is interrupted by a predicted short 32bp intron, represented by a thin white line within the gray ORF). Thick dotted lines represent target-site duplication (TSDs). Black thin bars represent transcripts either deduced from *S. mansoni* high-throughput EST sequencing evidence (for Merlin_Sm1) or directly cloned (for SmTRC1). Light gray bars indicate predicted transcripts or regions of transcripts, based on *S. japonicum* EST data (for Merlin_Sm1) or *S. mansoni* GenBank EST data (for SmTRC1). Thin lines connecting bars indicate intronic regions.

conservation of several amino acids in three different regions (Fig. 3) of the transposase_21 domain (PFAM 02992), a characteristic motif that is present in CACTA transposons. In two organisms, *Aedes aegypti* and *Tribolium castaneum*, a full transposon sequence was detected including the TIRs and it is possible to identify a 2 bp direct repeat flanking these TIRs, indicating a TSD event.[24] Using the *A. aegypti* TRC1 deduced protein as query in a TBLASTN search against GenBank we have now found a new complete 3,969 bp long TRC transposon in another mosquito species, *Anopheles gambiae*, with a 17 bp TIR flanking the transposon and a 2 bp long TSD. All the described animal full-length TRC transposons contain a core CCC sequence in their TIRs, similarly to the CACTA core described in CACTA transposons. Two TRC elements have been found in *A. gambiae* and *A. aegypti*, which are vectors of malaria and yellow fever, respectively, and represent novel transposon families in mosquitoes in addition to those described in a companion chapter in this book. Only one transcript sequence with considerable similarity with SmTRC1 was found in *S. japonicum* sequences at public databases; this transcript has some similarity (E-value 10^{-20}) at the protein level[24] (AAW24935.1), but no detectable similarity at the nucleotide level (AY813203.1), suggesting the existence of a TRC element in *S. japonicum*.

RT-PCR amplification experiments using primers designed from ESTs mapping to the SmTRC1 element, followed by cloning and sequencing resulted in the characterization of a number of diverse transcripts generated by SmTRC1.[24] Three transcripts encoding different protein products have been characterized (Fig. 2). *PrA* and *PrB* transcripts do not overlap the transposon ORF, and the proteins encoded by these transcripts do not have a detectable similarity to any described protein. Transcript *PrC* includes a region at the 3' end of the transposon ORF and was not fully characterized, lacking a region of its 5' end (Fig. 2, black

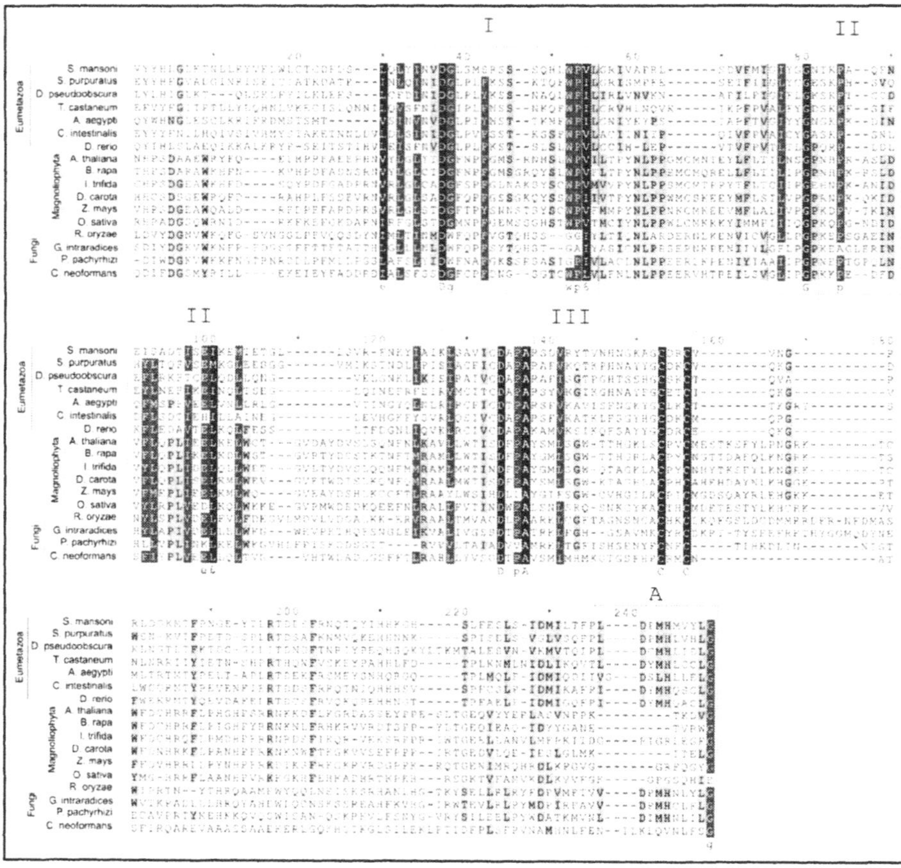

Figure 3. Multiple alignment of the Transposase_21 domains of proteins of CACTA related DNA transposons from diverse organisms. Typical plant CACTA transposon sequences from six Magnoliophyta were included in the alignment. In addition, eleven novel CACTA-related elements were included: seven from Eumetazoa and four from Fungi. Shading indicates the level of conservation of each residue. Boxes with Roman numbers I to III indicate conserved motifs of the Transposase_21 domain in all organisms. Box marked with A indicates a Transposase_21 motif displayed only by Eumetazoa and Fungi proteins. Reproduced with permission from DeMarco R, Venancio TM, Verjovski-Almeida S. BMC Evol Biol 2006; 6:89. ©2006 DeMarco et al.

bar). Additionally, one *S. mansoni* EST from the public database overlaps to the beginning of the ORF (Fig. 2, *PrC* light gray bar) and we hypothesize that this EST may represent the 5' end of PrC. This hypothetical transcript would cover practically the whole transposon ORF (Fig. 2, gray bar), encoding a 407 amino acids long protein that includes the Transposase_21 domain. The overall transcript structure is similar to that seen in the En/Spm transposon, which leads to the hypothesis that SmTRC-PrA may perform an analogous role to the DNA binding function described for TNPA in plant transposons,[31,32] although no sequence similarity could be detected. SmTRC-PrC may perform an analogous function to the longer protein TPND of En/Spm, since both proteins display the Transposase_21 domain. Interestingly, we also found several spliced ESTs in GenBank mapping the *A. gambiae* transposon and suggesting that all eumetazoa TRCs may generate alternatively spliced transcripts, in a similar manner as SmTRC1.

Application of Schistosome DNA Transposons to Transgenesis and Mutagenesis

The current release of the transcriptome and genome information of schistosomes[16,18,19] increased by several fold the number of described genes, including many with undetermined function for which no similarity to previously described proteins was detected. In this context, it is extremely desirable to have new tools that will permit transgenesis and mutagenesis of schistosomes, which could provide insights into the function and regulation of *Schistosoma* genes. Some pioneering work has been recently performed to introduce transgenes in schistosomes, including the use of vertebrate retrovirus[33] and ballistics,[34,35] with varying degrees of success.

DNA transposons are the most frequently used mobile elements for manipulating and transforming genomes of prokaryotic and eukaryotic organisms[36] and examples of its successful use in invertebrate and vertebrate transgene or mutagenesis experiments are abundant.[37-40] Both Merlin_Sm1 and SmTRC1 schistosome transposons display copies with a high level of integrity, with long and apparently intact ORFs that would encode active transposases, and intact TIRs and TSDs, suggesting recent transposition of these elements. Therefore, it is possible to speculate that both elements are potential tools to manipulate the genome of schistosomes and other platyhelminths, given the potential intrinsic mobility associated to them.

Apparently, Merlin_Sm1 requires only one protein to transpose, which would permit the construction of a two-component DNA transposon delivery system, dissociating the gene responsible for mobility from the DNA to be delivered, as employed for other mobile elements. The use of such design for construction of a DNA transposon delivery system would enhance the chromosomal stability of a transgene, since in principle it provides the means to induce and control the source of transposase.[36]

In contrast, SmTRC1 would require more than one protein to transpose, similar to what has been described for CACTA transposons.[31,32] Nevertheless, it has been shown that the characteristics of CACTA elements permit their use in insertional mutagenesis in plants[41,42] and it is possible that SmTRC1 may display a similar behavior, making it an interesting new tool for insertional mutagenesis.

Acknowledgements

The work in the authors' laboratory is supported by a grant from FAPESP, Fundação de Amparo a Pesquisa do Estado de São Paulo, and by fellowships from FAPESP; CAPES, Coordenação de Aperfeiçoamento de Pessoal de Nível Superior, and CNPq, Conselho Nacional de Desenvolvimento Científico e Tecnológico, Brasil.

References

1. SanMiguel P, Gaut BS, Tikhonov A et al. The paleontology of intergene retrotransposons of maize. Nat Genet 1998; 20(1):43-45.
2. Messing J, Dooner HK. Organization and variability of the maize genome. Curr Opin Plant Biol 2006; 9(2):157-163.
3. Charlesworth B, Sniegowski P, Stephan W. The evolutionary dynamics of repetitive DNA in eukaryotes. Nature 1994; 371(6494):215-220.
4. Kidwell MG, Lisch D. Transposable elements as sources of variation in animals and plants. Proc Natl Acad Sci USA 1997; 94(15):7704-7711.
5. Ganko EW, Bhattacharjee V, Schliekelman P et al. Evidence for the contribution of LTR retrotransposons to C. elegans gene evolution. Mol Biol Evol 2003; 20(11):1925-1931.
6. Cordaux R, Udit S, Batzer MA et al. Birth of a chimeric primate gene by capture of the transposase gene from a mobile element. Proc Natl Acad Sci USA 2006; 103(21):8101-8106.
7. Morrish TA, Gilbert N, Myers JS et al. DNA repair mediated by endonuclease-independent LINE-1 retrotransposition. Nat Genet 2002; 31(2):159-165.
8. Casacuberta E, Pardue ML. HeT-A elements in Drosophila virilis: Retrotransposon telomeres are conserved across the Drosophila genus. Proc Natl Acad Sci USA 2003; 100(24):14091-14096.

9. Haren L, Ton-Hoang B, Chandler M. Integrating DNA: Transposases and retroviral integrases. Annu Rev Microbiol 1999; 53:245-281.
10. Capy P, Bazin C, Higuet D et al, eds. Dynamics and Evolution of Transposable Elements. Austin/New York: Landes Bioscience/North American Distributor Chapman and Hall, 1998.
11. Craig NL, Craigie R, Gellert M et al, eds. Mobile DNA II. Washington, DC: ASM Press, 2002.
12. Feschotte C. Merlin, a new superfamily of DNA transposons identified in diverse animal genomes and related to bacterial IS1016 insertion sequences. Mol Biol Evol 2004; 21(9):1769-1780.
13. WHO. TDR strategic direction for research: Schistosomiasis. Geneve: World Health Organization, 2002.
14. Gryseels B, Polman K, Clerinx J et al. Human schistosomiasis. Lancet 2006; 368(9541):1106-1118.
15. Simpson AJ, Sher A, McCutchan TF. The genome of Schistosoma mansoni: Isolation of DNA, its size, bases and repetitive sequences. Mol Biochem Parasitol 1982; 6(2):125-137.
16. Wilson RA, Ashton PD, Braschi S et al. 'Oming in on schistosomes: Prospects and limitations for post-genomics. Trends Parasitol 2007; 23(1):14-20.
17. Brindley PJ, Laha T, McManus DP et al. Mobile genetic elements colonizing the genomes of metazoan parasites. Trends Parasitol 2003; 19(2):79-87.
18. Verjovski-Almeida S, DeMarco R, Martins EA et al. Transcriptome analysis of the acoelomate human parasite Schistosoma mansoni. Nat Genet 2003; 35(2):148-157.
19. Hu W, Yan Q, Shen DK et al. Evolutionary and biomedical implications of a Schistosoma japonicum complementary DNA resource. Nat Genet 2003; 35(2):139-147.
20. DeMarco R, Kowaltowski AT, Machado AA et al. Saci-1, -2 and -3 and Perere, four novel retrotransposons with high transcriptional activities from the human parasite Schistosoma mansoni. J Virol 2004; 78:2967-2978.
21. DeMarco R, Machado AA, Bisson-Filho AW et al. Identification of 18 new transcribed retrotransposons in Schistosoma mansoni. Biochem Biophys Res Commun 2005; 333(1):230-240.
22. Laha T, Brindley PJ, Verity CK et al. Pido, a non-long terminal repeat retrotransposon of the chicken repeat 1 family from the genome of the Oriental blood fluke, Schistosoma japonicum. Gene 2002; 284(1-2):149-159.
23. Copeland CS, Lewis FA, Brindley PJ. Identification of the Boudicca and Sinbad retrotransposons in the genome of the human blood fluke Schistosoma haematobium. Mem Inst Oswaldo Cruz 2006; 101(5):565-571.
24. DeMarco R, Venancio TM, Verjovski-Almeida S. SmTRC1, a novel Schistosoma mansoni DNA transposon, discloses new families of animal and fungi transposons belonging to the CACTA superfamily. BMC Evol Biol 2006; 6:89.
25. Hua-Van A, Le Rouzic A, Maisonhaute C et al. Abundance, distribution and dynamics of retrotransposable elements and transposons: Similarities and differences. Cytogenet Genome Res 2005; 110(1-4):426-440.
26. Laski FA, Rio DC, Rubin GM. Tissue specificity of Drosophila P element transposition is regulated at the level of mRNA splicing. Cell 1986; 44(1):7-19.
27. Evgen'ev MB, Zelentsova H, Shostak N et al. Penelope, a new family of transposable elements and its possible role in hybrid dysgenesis in Drosophila virilis. Proc Natl Acad Sci USA 1997; 94(1):196-201.
28. McClintock B. The origin and behavior of mutable loci in maize. Proc Natl Acad Sci USA 1950; 36(6):344-355.
29. Pereira A, Cuypers H, Gierl A et al. Molecular analysis of the En/Spm transposable element system of Zea mays. EMBO J 1986; 5(5):835-841.
30. Masson P, Rutherford G, Banks JA et al. Essential large transcripts of the maize Spm transposable element are generated by alternative splicing. Cell 1989; 58(4):755-765.
31. Frey M, Reinecke J, Grant S et al. Excision of the En/Spm transposable element of Zea mays requires two element-encoded proteins. EMBO J 1990; 9(12):4037-4044.
32. Masson P, Strem M, Fedoroff N. The tnpA and tnpD gene products of the Spm element are required for transposition in tobacco. Plant Cell 1991; 3(1):73-85.
33. Kines KJ, Mann VH, Morales ME et al. Transduction of Schistosoma mansoni by vesicular stomatitis virus glycoprotein-pseudotyped Moloney murine leukemia retrovirus. Exp Parasitol 2006; 112(4):209-220.
34. Wippersteg V, Kapp K, Kunz W et al. HSP70-controlled GFP expression in transiently transformed schistosomes. Mol Biochem Parasitol 2002; 120(1):141-150.
35. Osman A, Niles EG, Verjovski-Almeida S et al. Schistosoma mansoni TGF-beta receptor II: Role in host ligand-induced regulation of a schistosome target gene. PLoS Pathog 2006; 2(6):e54.
36. Feschotte C. The piggyBac transposon holds promise for human gene therapy. Proc Natl Acad Sci USA 2006; 103(41):14981-14982.

37. Handler AM. A current perspective on insect gene transformation. Insect Biochem Mol Biol 2001; 31(2):111-128.
38. Dupuy AJ, Clark K, Carlson CM et al. Mammalian germ-line transgenesis by transposition. Proc Natl Acad Sci USA 2002; 99(7):4495-4499.
39. Miskey C, Izsvak Z, Plasterk RH et al. The frog prince: A reconstructed transposon from Rana pipiens with high transpositional activity in vertebrate cells. Nucleic Acids Res 2003; 31(23):6873-6881.
40. Izsvak Z, Ivics Z. Sleeping beauty transposition: Biology and applications for molecular therapy. Mol Ther 2004; 9(2):147-156.
41. Wisman E, Cardon GH, Fransz P et al. The behaviour of the autonomous maize transposable element En/Spm in Arabidopsis thaliana allows efficient mutagenesis. Plant Mol Biol 1998; 37(6):989-999.
42. Tissier AF, Marillonnet S, Klimyuk V et al. Multiple independent defective suppressor-mutator transposon insertions in Arabidopsis: A tool for functional genomics. Plant Cell 1999; 11(10):1841-1852.

CHAPTER 9

Mobile Genetic Elements Resident in Hookworm Genomes

Thewarach Laha, Alex Loukas and Paul J. Brindley*

Abstract

Hookworms are intestinal nematode parasites of humans and other mammals, and are responsible for considerable morbidity in populations living in many developing countries. Several mobile genetic elements have been characterized from hookworm genomes. The structure and phylogenetic relationships of these characterized elements is reviewed here. The genome of dog hookworm *Ancylostoma caninum*, a model for the human hookworms *A. duodenale* and *Necator americanus*, includes at least two transposons, ***bandit*** and a ***transib*** element, and non-LTR retrotransposons known as ***dingo*** elements. The ***bandit*** transposon is a ***mariner***-like element, the ***transib*** element is related to ***transib***-elements involved in V(D)J recombination in vertebrate adaptive immunity, and the ***dingo*** elements are clade RTE non-LTR retrotransposons. It is likely that numerous other mobile genetic elements have colonized the hookworm genome. We anticipate that the information of these elements reviewed here from *A. caninum* will provide a bridgehead for research into the larger complement of mobile genetic elements and other repetitive sequences that have colonized the genomes of the human hookworm species and other hookworms.

Introduction

Almost one billion people throughout tropical and sub-tropical latitudes are infected with hookworms. In the countries affected, hookworm infection is often the major contributor to iron-deficiency anemia, a direct consequence of the blood-feeding activities of hookworms.[1] Comparatively little is known about the genome or population genetics of hookworms. The *Ancylostoma caninum* hookworm of canids is a model for studying human infections and information from its genome coupled with functional genomics and proteomics can accelerate progress towards hookworm control. The karyotype of only one hookworm species, the dog hookworm, *A. caninum*, is known where $n = 6$.[2] Hookworms are dioecious and sex determination is by an XX-XO mechanism as in their free-living relative, the model nematode *Caenorhabditis elegans*.[3] The genome size was estimated by flow cytometry of isolated nuclei to be 347 ± 1.2 Mb, substantially larger than other Rhabditina species (Mitreva et al, 2008). By contrast, flow cytometric based techniques have shown that the haploid genome size of two trichostrongyle nematodes, *Haemonchus contortus* and *Teladorsagia circumcincta*, is ~50 Mb in length.[4] Trichostrongyle nematodes are more closely related to hookworms than is the free-living nematode, *C. elegans*.[5] The larger size of the genome of the hookworm may be the result of considerable colonization and/or expansion by mobile genetic elements, as is known from other groups of closely related species.[6]

*Corresponding Author: Paul J. Brindley—Department of Microbiology, Immunology and Tropical Medicine, George Washington University Medical Center, Ross Hall, Room 448, 2300 Eye Street, NW, Washington, DC 20037 USA. Email: mtmpjb@gwumc.edu

Mobile Genetic Elements in Metazoan Parasites, edited by Paul J. Brindley.

As a step towards a full-scale *A. caninum* genome project, Mitreva and coworkers recently generated 104,000 genome survey sequences (GSSs).[7] Gene finding identified 5,538 genes in the GSS assembly, for a total of 9,113 nonredundant *A. caninum* genes when EST sequences were also considered. Over 20,000 ESTs from *A. caninum* and the related parasite, *A. ceylanicum*, have been characterized to some degree,[8-10] including transcripts from the gut of adult worms.[11] Interestingly, many of the genes share homologues with *C. elegans*, highlighting the suitability of this free-living nematode as a model for hookworm developmental biology.[10] At present, the GSS database at GenBank contains more than 100,000 genome survey sequences (GSS) from *A. caninum*, which when assembled provide 57.6 Mb unique sequences, establishing a tractable framework for an eventual genome sequence. It can be anticipated that an enhanced understanding of the hookworm genome will aid in the control of hookworm disease and hookworm-associated anemia, including the development of new antiparasite interventions.[12]

A substantial proportion of the genome of most metazoans is composed of repetitive sequences, including various types of mobile genetic elements (MGEs). MGEs are drivers of genome evolution.[13] In addition to this role, from a practical perspective MGEs offer potential as transgenesis and gene silencing vectors,[14-16] technologies that have yet to be reliably established for the study of parasitic nematodes. Problematically, however, their interspersed, repetitive nature can impede progress during genome sequencing using shotgun sequencing approaches through the confounding effects of their repetitions on sequence assembly algorithms.[17-18] For these and other reasons, knowledge of hookworm MGEs is of theoretical and practical value.

We have reported the presence of a family of nonlong terminal repeat (LTR) retrotransposons, the *dingo* retrotransposons, from the genome of *A. caninum*[19] and the presence of a *mariner* like transposon, termed *bandit*, within the genome of *A. caninum*.[20] *Bandit* is a DD(34)D family *mariner*-like transposon which, intriguingly, is much more closely related to the human *mariner*-like element *Hsmar1* than to any other MLE so far reported from other species of the phylum Nematoda. In addition, bioinformatics approaches have identified (at least partial fragments) of a *Transib*-like transposon in the genome of *A. caninum*.[21] In depth descriptions of these several mobile genetic elements from hookworm genomes are provided below.

dingo: Isolation of the *dingo* Retrotransposons

Members of the RTE (retrotransposable element) clade of nonlong terminal repeat (LTR) retrotransposon are widely distributed among eukaryote taxa, with representatives known from *C. elegans*, mammals, mosquitoes, schistosomes and other taxa. An RTE retrotransposon had not, however, been characterized in detail from a parasitic nematode until the report of the *dingo* elements. Laha and coworkers described two discrete copies of an RTE-like non-LTR retrotransposon from the genome of the dog hookworm *A. caninum*, and named the elements *dingo-1* and *dingo-2*.[19]

The *dingo* elements were assembled as contigs from cloned fragments of genomic DNA and from GSS in GenBank.[19] First, a cloned genomic fragment termed H012, a 1. 5 kb *Hind* III fragment, was isolated by screening size selected partial *A. caninum* genomic DNA libraries with retrotransposon like gene probes, specifically sequences of GenBank accessions AW588541 and BQ666481. Subsequently, using H012 as the nucleation point, a contig was assembled from the GSS of *A. caninum* retrieved from the public databases, as illustrated in Figure 1. The contig encoded the predicted sequence of a new non-LTR retrotransposon-like element, which was termed *dingo-1*, and shared structural similarities to other retrotransposons characterized previously from other helminths including *C. elegans* and *S. mansoni*.[22,23] A second, related retrotransposon sequence, termed *dingo-2*, was assembled in the same fashion as for the *dingo-1* consensus sequence. For this second assembly, the sequence of the AcRTE2 probe (BQ666481) represented the nucleation point from which the contig was assembled using GSS of *A. caninum* retrieved from GenBank.

Structural Features of the *dingo* Retrotransposons

These two hookworm elements are discrete copies of a retrotransposon that we termed, *dingo*. The dingo is the wild dog of Australia, a secretive, peripatetic carnivore introduced to

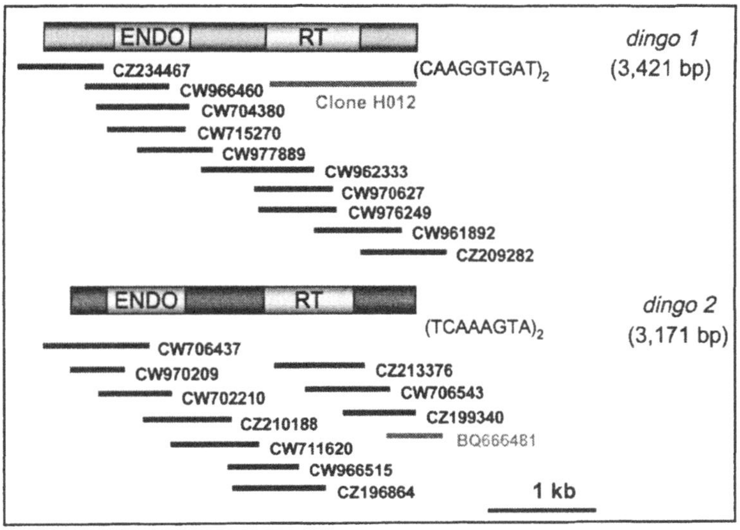

Figure 1. Schematic diagrams of contig assembly for, and the structure of, the *dingo-1* and *dingo-2* non-LTR retrotransposons from the genome of the hookworm, *Ancylostoma caninum*. The domain structure of the *dingo* elements is presented, along with the positions, names and lengths of clones or GenBank accessions used to assemble the contigs. RT, reverse transcriptase; EN, apurinic-apyrimidinic endonuclease. Adapted from: Laha T et al. Exp Parasitol 2006; 113:142-153;[19] ©2006 with permission from Elsevier.

the Australian continent reasonably early in human history. We considered the movement of the new *dingo* MGE to exhibit some analogous characteristics to this wild canine including movement through the hookworm parasite genome and to be an ancient resident of the hookworm genome. In addition, just as dingoes are naturally parasitized by *A. caninum* (see ref. 19), the *dingo* retrotransposon is a genetic parasite of the hookworm genome. The two discrete copies are termed *dingo-1* and *dingo-2* (Fig. 1). *dingo* includes a single open reading frame (ORF) that encodes a polyprotein constituted of endonuclease (EN) and reverse transcriptase (RT) domains, in that order. Like many other non-LTR retrotransposons, the 3'-ends of *dingo* terminate with short repeat units; *dingo-1* terminates with the short terminal repeat of the nonamer CAAGGTGAT whereas *dingo-2* ends with a repeat of the octamer TCAAAGTA. The full-length *dingo-1* element is 3,421 bp in length, with a predicted ORF of 919 deduced amino acid (aa) residues. The EN domain of *dingo-1* was 239 aa and the RT domain about 274 aa in length. *dingo-2* was predicted to be 3,171 bp in length, with an ORF of about 965 aa residues. The EN and RT domains of *dingo-2* are 242 aa and 272 aa long, respectively. *dingo*1 has a long 5'-UTR (567 bp) whereas the 5'-UTR of *dingo-2*, of 231 bp, is shorter. Their 3'-UTRs are similar in length; 94 bp and 68 bp for *dingo-1* and *-2*, respectively. *dingo-1* and *dingo-2* exhibit 40% amino acid residue identity and 58% similarity across their entire ORFs.

dingo Is a Novel Non-LTR Retrotransposon from the RTE Family

Sequence comparisons by Blastx revealed that *dingo* was similar to *RTE1* from *C. elegans*. *dingo-1* shared 42% identity and 51% similarity to RTE1 from *C. elegans* and 27% identity and 45% similarity to *SR2* from *S. mansoni*.[24,25] The predicted RT domains of the *dingo* elements were aligned with orthologous domains of numerous other non-LTR retrotransposons including representatives from 11 clades of non-LTR retrotransposons, as defined by Eickbush and colleagues,[26] and with RT-like sequences of Group II introns serving as the outgroup.[27,28] Phylogenetic comparison of the RT domains of these diverse elements revealed that *dingo-1* and *dingo-2* were closely related, and that the closest relative of *dingo* was *RTE-1* from *C.*

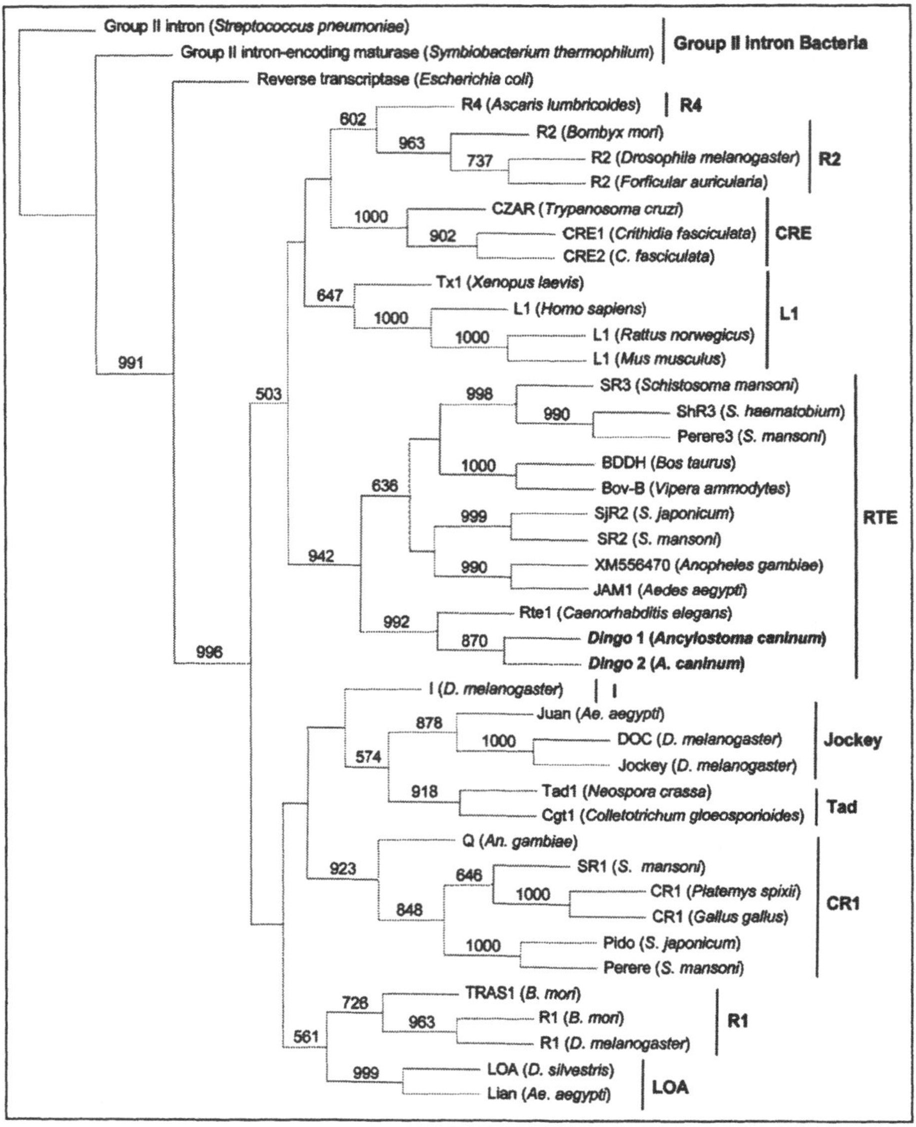

Figure 2. Phylogram constructed using the neighbor-joining method to compare the relationships among reverse transcriptases of the *dingo* retrotransposons and of representative elements belonging to the major clades of non-LTR retrotransposons from a range of host genomes. Representatives of 10 clades of non-LTR retrotransposons including the RTE, L1, CRE, R1 and R4 clades were included in the analysis. Bootstrap values, where 500 or greater from a maximum of 1,000 replicates, are presented at the nodes. Adapted from: Laha T et al. Exp Parasitol 2006; 113:142-153;[19] ©2006 with permission from Elsevier.

elegans. Furthermore, *dingo* was closely related to the RTE clade non-LTR retrotransposons including the *SR2*- and *SR3*-like retrotransposons from schistosomes, the JAM1-like elements from mosquito genomes, and the BDDH and Bov-B elements from cattle and vipers. The RTE elements at large were related to the L1, CRE, R1 and R2 clade retrotransposons, and more distantly related to other LTR retrotransposon clades including CR1 and Jockey (Fig. 2).

Laha and cowokers also constructed a phylogenetic tree based on the EN domains of seven clades of non-LTR retrotransposons, with bacterial endonucleases as the outgroup.[19] The topography of the EN tree, and the position of *dingo* within the RTE clade, was similar to the topography represented on the RT-based tree, confirming the inclusion of *dingo* as an RTE clade element. Interestingly, in the EN tree, *dingo-1* was more closely related to *RTE-1* of *C. elegans* than to *dingo-2 of A. caninum*. Nonetheless, the phylogenetic relationships revealed using both the RT and EN domains demonstrated conclusively that *dingo* was an RTE-like non-LTR retrotransposon.

Numerous Copies of *dingo* Interspersed throughout the *A. caninum* Genome

Dot blot hybridization indicated that as many as 100-1,000 copies of *dingo-1* reside within the genome of *A. caninum*, while detection by RT-PCR of transcripts encoding *dingo*-like elements suggested that *dingo-1* and *-2* are retrotranspositionally active within the genome of *A. caninum*. To investigate the nature of integration sites or target sequences of the new retrotransposon within the hookworm genome, sequences flanking of the 5'-and 3'-termini of several copies of *dingo- 1* and *-2* were employed as queries to search the GSS and EST entries for *A. caninum* in GenBank. These searches revealed no significant matches to any sequences encoding genes of *Ancylostoma* species. However, they did reveal that *dingo* elements appear to target AT-rich sites, a similar preference to *L1* retrotransposons within the human genome.[29] More specifically, at least four copies of *dingo* (CZ198614, CW706437, CZ248710 and CW713425) have integrated into microsatellite-like $(TA)_n$ loci (Fig. 6A,B). Whereas target site specificity does not appear to be stringent for *dingo*, it can be expected to reflect the recognition sequence of the *dingo* endonucleases.[26-29]

Transposon *bandit*

Laha and coworkers reported recently that a *mariner*-like transposon termed *bandit* has colonized the genome of the hookworm *A. caninum*.[20] Although MLEs exhibit a broad host range, and are identified in other nematodes, the closest phylogenetic relative of *bandit* is the *HsMar*1 element of humans. This surprising finding suggests that *bandit* was transferred horizontally between hookworm parasites and their mammalian hosts. The novel *mariner*-like element (MLE) *bandit* was characterized from the genome of the dog hookworm, *Ancylostoma caninum*. The consensus sequence of the *bandit* transposon was 1,285 bp in length. The new transposon was flanked by perfect terminal inverted repeats of 32 nucleotides in length with a common target site duplication TA, and it encoded an open reading frame of 342 deduced amino acid residues. The uninterrupted transposase ORF of *bandit* was amplified from RNA, indicating that this transposon is actively transcribed in hookworms.

Discovery of *bandit* in the Genome of *A. caninum*

A positive clone was identified from the *A. caninum* genomic DNA libraries that was screened with a reverse transcriptase-like gene probe, clone H118 (GenBank DQ377715).[20] The clone showed sequence identity with *mariner*-like transposons from many eukaryotes including *mariner* from *Homo sapiens* and *mariner* from *Bos taurus*. Sequence analysis revealed that clone H118 contained sequence which encoded part of a transposase protein. The consensus full length transposon was constructed using clone H118 and multiple GSSs identified by homology searches from the GenBank database (GenBank accession numbers CW709686, CZ213904 and CZ241797) (Fig. 3). We termed the new transposon *bandit*, in keeping with the informal convention of naming mobile genetic elements with terms suggestive of a peripatetic lifestyle (e.g., *mariner*, *hobo* and *fugitive*). The name *bandit* seems appropriate since a bandit is difficult to apprehend, and in this present context, it appears that *bandit* has moved stealthily between hookworms and their mammalian hosts.

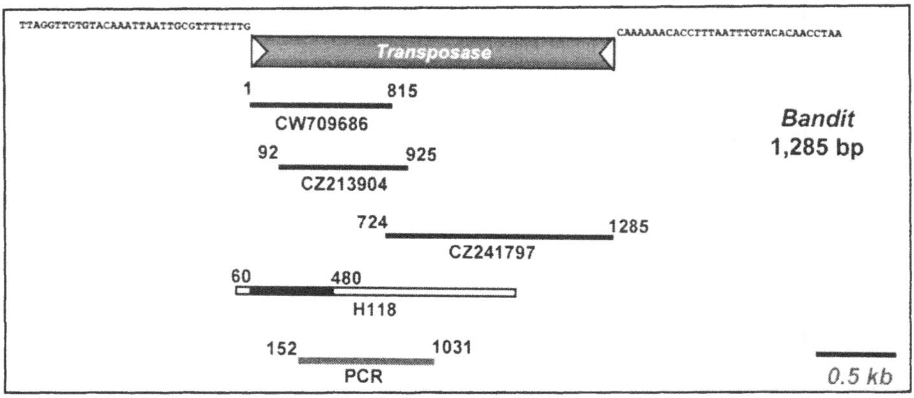

Figure 3. Schematic diagram representing clones and genomic survey sequences (GSS) from public databases which were used to assemble the consensus sequence of the *bandit* transposon from the genome of *Ancylostoma caninum*. Numbers on clones represent the nucleotide positions within the consensus, full length *bandit* sequence. GenBank accession numbers of contributing GSS clones are provided. The sequences of the terminal inverted repeats are presented in the top panel. In clone H118, the black colored region is bandit sequence whereas the white region on nonbandit encoding DNA. Adapted from: Laha T et al. PLoS Negl Trop Dis 2007; 1(1):e35.[20]

Structural Features of the *bandit* Transposon

The consensus sequence of *bandit* was 1,285 bp flanked by 32 nt perfect terminal inverted repeats at each extremity with a common target site duplication TA (Fig. 3). *bandit* has one ORF of 342 amino acid residues encoding for a transposase enzyme. The *bandit* transposase contained the conserved DD34D motif that is found in the active site of the catalytic C-terminal domain of *mariner*-like transposons as opposed to the DDE motif found in the *Tc1*-like elements.[30,31] The ORF of the *bandit* showed highest similarities to *Hsmar1* from human (55% identity, 70% similarity), *Bos taurus* (54% identity, 70% similarity) and *Tc1* of *C. elegans* (41% identity, 58% similarity), *HcTc*1 of *Haemonchus contortus* (22% identity, 42% similarity). The perfect inverted repeats of 32 bp are the standard length for *mariner*-like elements compared with 54 bp for *Tc1* from *C. elegans* and 55 bp for *HcTc*1 from *H. contortus*.[31-33] In addition to the catalytic triad, *bandit* contains most of the additional canonical features of *mariner*-like elements (MLEs); the WVPHEL motif (RVPHEL in *bandit*) and YSPDLAP (CSPDLSP in *bandit*).[34] However, *bandit* does not contain the conserved FLHDNARPH motif that overlaps the second D of catalytic triad in most MLE transposases. In *bandit*, this motif is replaced by a LLHDNARSH motif.[35-36]

bandit is a Novel *mariner*-like Transposon of the Cecropia Subfamily

Phylogenetic comparisons confirmed that the ORF encoded a *mariner*-like transposase, which included conserved catalytic domains, and that the *bandit* transposon belonged to the *cecropia* subfamily of MLEs. A phylogenetic tree was constructed based on the sequence alignment of the entire transposase ORFs of *bandit* and 37 other transposon sequences available in public databases. A neighbor-joining tree with 1,000 replicates revealed that *bandit* is most closely related to *Hsmar*1 from *Homo sapiens* (Fig. 4). *Mariner*-like transposons can be classified into six subfamilies.[37-38] *Bandit* forms a clade with elements from the cecropia subfamily (Fig. 4), and this diphyletic clade

Figure 4, viewed on next page. Phylogram constructed using the neighbor-joining method to compare the relationships among transposases of the *bandit* transposon and of representative mariner-like elements belonging to the *Tc1/mariner* superfamily of transposons from a range of host genomes. Representatives of six clades of mariner-like elements including the mori, irritans, mauritiana, and cecropia were included in the analysis. Legend continued on following page.

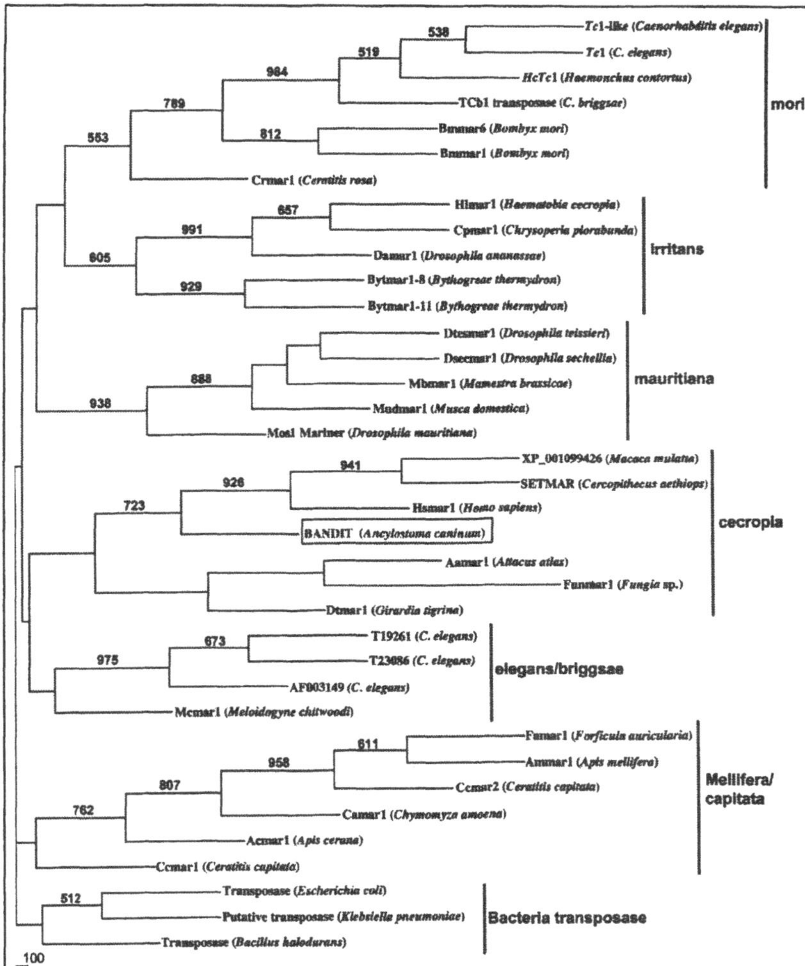

Figure 4, legend continued from previous page. The elements used in the tree includes Tc1-like (AAD12818) and Tc1 (P03934), T19261, T23086 and AF003149 from *C. elegans*, HcTc1 (AAD34306) from *Haemonchus contortus*, TCb1 (CAA30681) from *C. briggsae*, Bmmar1 (U47917) and BmMar6 (AAN06610) from *Bombyx mori*, Crmar1 (AAK61417) from *Ceratitis rosa*, Himar1 (ABB59013) mutagenesis vector pFNLTP16H3, Cpmar1 (AAC46945) from *Chrysoperla plorabunda*, Damar1 (DAU11648) from *Drosophila ananassae*, Bytmar1-8 (CAD45868) and Bytmar1-11 (CAD45369) from *Bythogreae thermydron*, Dtesmar1 (AAC28261) from *D. teissieri*, Dsecmar1 (AAC16609) from *D. sechellia*, Mbmar1 (AAL69970) from *Mamesta brassicae*, Mudmar1 (AK54758) from *Musca domestica*, Mos1 (pdb2F7T) from *D. mauritiana*, XP_001099426 from *Macaca mulatta*, SETMAR (ABC72092) from *Cercopithecus aethiops*, Hsmar1 (AAC52010) from *Homo sapiens*, Aamar1 (BAA21826) from *Attacus atlas*, Funmar1 (BAB32436) from Fungia sp., Dtmar1 (CAA50801) from *Girardia tigrina*, Mcmar1 (CAD26968) from *Meloidogyne chitwoodi*, Famar1 (AAO12863) from *Forficula auricularia*, Ammar1 (AAO12861) from *Apis mellifera*, Ccmar2 (AAO12864) from *Ceratitis capitata*, Camar1 (AAO12862) from *Chymomyza amoena*, Acmar1 (BAB86288) from *Apis cerana*, Ccmar1 (AAB17945) from *Ceratitis capitata*. The outgroup included transposases from Gram-positive and negative bacteria including *Bacillus halodurans* (BAA75315), *Escherichia coli* (AAB28848) and *Klebsiella pneumoniae* (CAB82575). Bootstrap values, where 500 or greater from a maximum of 1,000 replicates, are presented at the nodes. Adapted from: Laha T et al. PLoS Negl Trop Dis 2007; 1(1):e35.[20]

includes a branch of *bandit* and three primate-originated MLEs and a branch with *Funmar*1 from the coral *Fungia* species, *Aamar*1 from the atlas moth, *Attacus atlas* and *Dtmar*1 from the planarian, *Girardia tigrina*. The phylogenetic distance between human and hookworm is far greater than that reflected in the phylogenetic analysis of these transposons, suggesting that *bandit* is only distantly related to MLEs from nematodes that are closely related to *A. caninum*, and is much more similar to transposons from the mammals that are parasitized by the hookworm. For example, the MLE HcTc1 from the trichostrongyle parasite, *H. contortus* (a close relative of *A. caninum*) belongs to the *mori* clade of MLEs (Fig. 4). The remarkable identity between *bandit* and the primate MLEs, *Hsmar1* and SETMAR, indicates that *bandit* has spread through horizontal transmission from host to parasite, or vice versa.[20]

Close Identity of Hookworm and Human Transposons Confirmed by Homology Models

Homology models based on the crystal structure of *Mos1* revealed closer identity in active site residues of the catalytic domain including Ser281, Lys289 and Asp293 between *bandit* and *Hsmar1* than between *Mos1* and either *bandit* or *Hsmar1*. More specifically, the catalytic C-terminal domain of the predicted transpose ORF of *bandit* was modeled on the crystal structure of the C-terminal catalytic domain (residues 126-345) of *mos1* transposase from *Drosophila mauritiana* (pdb accession 2f7tA). The structural alignment spanned residues 158-345 of *mos1* and 178-342 of *bandit*. The general fold of the *bandit* catalytic domain is highly conserved with that of *mos1*. The first alpha helix and beta sheet of the catalytic domain of *bandit*, including the first catalytic Asp residue, were too dissimilar to *mos1* to be included in the model; however, the rest of the domain revealed similar active site architecture. Because *bandit* is most similar to human *Hsmar1* at the primary sequence level, we also modeled the catalytic domain of *Hsmar1* transposase on the crystal structure of *mos1*.[39] The sequence conservation between *mos1* and *Hsmar1* also was high (Fig. 5). Surprisingly, when the key active site residues of the catalytic domains of *bandit* and *Hsmar1* were compared with those of *mos1*, we observed that *bandit* and *Hsmar1* had identical active site residues but, by contrast, three of these residues had nonconservative substitutions in *mos1* (Fig. 5).[20]

The *Transib* Transposon

The hookworm genome also includes elements with identity to the Transib superfamily of transposons. In vertebrates, the *Transib* transposon has mutated to form the RAG1 protein and recombination signal sequences involved in catalyzing B and T cell receptor gene V(D)J recombination.[21] *Transib* is known from GSS of *A. caninum* although the full length *Transib* element from *A. caninum* or indeed other hookworms has not yet been characterized. Nonetheless, Kapitonov and Jurka[21] used the hookworm sequences and related sequences from mosquitoes and other invertebrates to make the seminal observation that the RAG1 of lampreys and higher vertebrates evolved from the transposase of *Transib*.

Transib is one of about 17 superfamilies of DNA transposons known from eukaryotes.[21,40,41] Like other DNA transposons, the *Transib* transposons exist as autonomous and nonautonomous elements. Autonomous *Transib* transposons are 3-4 kb long and encode a ~700 amino acid transposase that is dissimilar to transposases from other superfamilies. Computational analysis of *Transib* elements, including numerous insertions into copies of other transposons, demonstrated that *Transib* transposons are flanked by 5-bp target site duplications (TSDs), which also distinguish this superfamily from all the others.[41,42] As with transposition in *mariner* and other transposons, *Transib* transpositions can be expected to be catalyzed by the binding of the transposase to terminal inverted repeats (TIRs) of autonomous and nonautonomous *Transib* elements.[21,42]

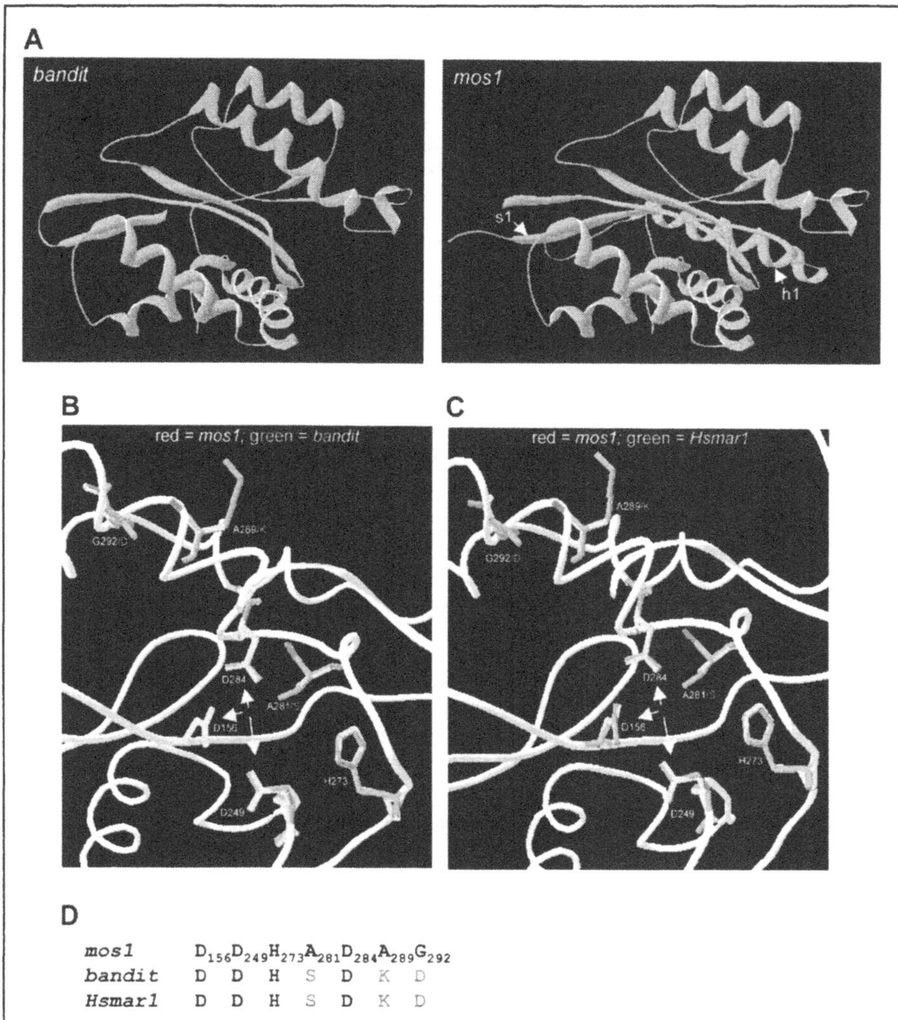

Figure 5. Homology models of *bandit* from *Ancylostoma caninum* and *Hsmar1* from *Homo sapiens* based on the crystal structure of the *mos1* element from *Drosophila mauritiana*. Ribbon diagram showing the predicted structure of the catalytic domains of *bandit* and *mos1* (A). s1 and h1 refer to β sheet number 1 and α helix number 1 of the *mos1* catalytic domain—homologous regions were present in *bandit* but were not included in the model. Superimposition of the catalytic active sites of *bandit* (B) and *Hsmar1* (C) on the crystal structure of *mos1* highlighting the residues involved in catalysis. Conserved active site residues are labeled in red font; where bandit or Hsmar1 active site residues differ from mos1, the substitution is denoted in green font. Yellow arrows denote the three catalytic Asp residues. Numbering of side chains is based on the *mos1* sequence. Comparison of the residues predicted to be involved in catalysis from *bandit, Hsmar1* and *mos1* (D). Residues selected were based on the crystal structure of *mos1*. A color version of this image is available online at www.landesbioscience.com/curie. Adapted from: Laha T et al. PLoS Negl Trop Dis 2007; 1(1):e35.[20]

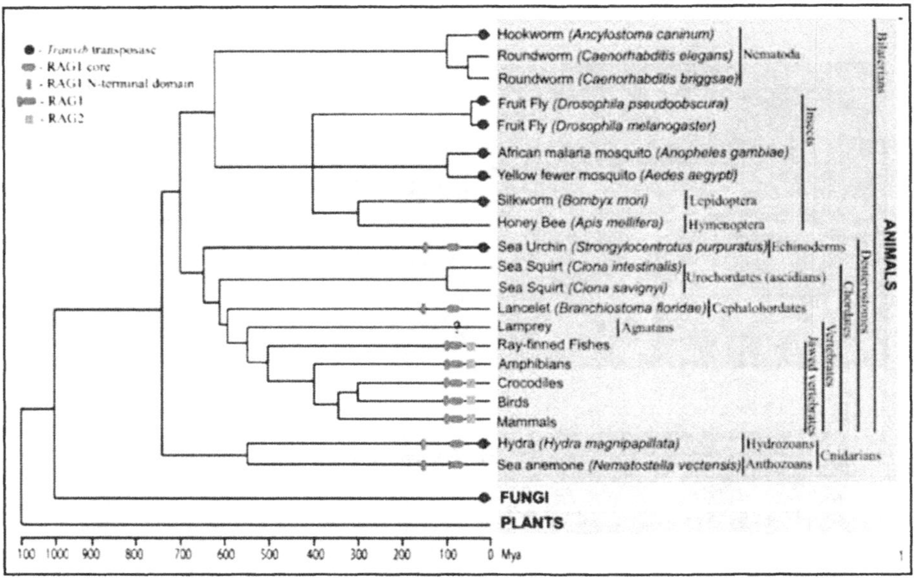

Figure 6. Schematic presentation of *Transib* transposons, RAG1, RAG2, and RAG1-Like proteins in eukaryotes. Red circles mark species in which *Transib* transposes have been identified. Gray squares indicate RAG2; orange and blue ellipses show the RAG1 core and RAG1 N-terminal domain, respectively. Lack of labels indicates that the *Transib* transposase and RAG1/2 are not present in the sequenced portions of the corresponding genomes. A color version of this image is available online at www.landesbioscience.com/curie. Figure and legend adapted from: Kapitonov VV, Jurka J. PLoS Biology 2005; 3(6):e181.[21]

In addition to the fruit fly (*Drosophila melanogaster*) and African malaria mosquito (*Anopheles gambiae*) genomes, in which *Transib* transposons were originally discovered, these genes are also present in diverse invertebrate phyla and in fungi, including the dog hookworm *A. caninum*, other species of fruit fly, mosquito (*Anopheles aegypti*, silk moth (*Bombyx mori*), red flour beetle (*Tribolium castaneum*), the freshwater flatworm *Schmidtea mediterranea*, hydra (*Hydra magnipapillata*), the sea urchin *Strongylocentrotus purpuratus*, and soybean rust, *Phakopsora pachyrhizi*. Recognizable *Transib* transposons have not been observed in plants or vertebrates.[21,43] It is intriguing that the *Transib* transposon has colonized the genome of a hookworm but it not known from the model nematodes *C. elegans* and *C. briggsae*. In like fashion, *Transib* is present in a mosquito genome but not in the honey bee (Fig. 6). In this regard, the distribution of the transposon conforms to the generalized capacity of cut-and paste transposons to move horizontally between taxons.

Concluding Remarks

Based on the genomes of *Caenorhabditis* species and several parasitic helminths including *Brugia malayi* and the human schistosomes,[44-46] it is likely that that the hookworm genome has been colonized not only by the *bandit* and *Transib* transposons and the *dingo* non-LTR retrotransposons, but also by numerous other waves of MGEs. Indeed, numerous other transcripts encoding reverse transcriptase are evident in the EST database of *A. caninum*, *A. ceylanicum* and *N. americanus* hookworms (http://nematode.net), indicating the presence of a substantial cohort of other endogenous retroviruses or retrotransposons. From a practical perspective, understanding of MGE complexity, diversity and copy numbers can be expected to facilitate the assembly and annotation of the hookworm genome sequence, a focus of current genome sequencing effort, http://nematode.net. As with other MGEs, endogenous hookworm transposon such as *bandit* and *Transib* hold potential as transgenesis vectors for manipulation of the hookworm genome, given the ability of other Tc1/mariner superfamily members such as *mos1* to transpose within the genomes of *C. elegans*, planarians and other species.[47] Moreover, *bandit* and *Transib* exhibit discrete target site preferences-AT rich and GC rich sites, respectively, so they may have complimentary roles. Finally, phylogenetic analysis of the MGEs of hookworms can be expected to provide further insights into the numerous episodes of commensalism and parasitism that are a hallmark of the Phylum Nematoda.[48,49]

References

1. Hotez PJ, Brindley PJ, Bethony JM et al. Helminth infections: the great neglected tropical diseases. J Clin Invest 2008; 118:1311-1321.
2. LeJambre LF, Georgi JR. Influence of fertilization on ovogenesis in Ancylostoma caninum. J Parasitol 1970; 56:131-137.
3. Blaxter M. Genes and genomes of Necator americanus and related hookworms. J Parasitol 2000; 30:347-355.
4. Leroy S, Duperray C, Morand S. Flow cytometry for parasite nematode genome size measurement. Mol Biochem Parasitol 2003; 128:91-93.
5. Holterman M, van der Wurff A, van den Elsen S et al. Phylum-wide analysis of SSU rDNA reveals deep phylogenetic relationships among nematodes and accelerated evolution toward crown Clades. Mol Biol Evol 2006; 23:1792-1800.
6. Petrov DA. Sangster TA, Johnston JS et al. Evidence for DNA loss as a determinant of genome size. Science 2000; 287:1060-1062.
7. Abubucker S, Martin J, Yin Y et al. The canine hookworm genome: analysis and classification of Ancylostoma caninum survey sequences. Mol Biochem Parasitol. 2008;157(2):187-92.
8. Daub J, Loukas A, Pritchard DL et al. A survey of genes expressed in adults of the human hookworm, Necator americanus. Parasitology 2000; 120:171-184.
9. Miranda RR, Costa-Junior LM, Campos AK et al. Identification of specific male and female genes in adult Ancylostoma caninum. Ann N Y Acad Sci 2004; 1026:199-202.
10. Mitreva M, McCarter JP, Arasu P et al. Investigating hookworm genomes by comparative analysis of two Ancylostoma species. BMC Genomics 2005; 6:58.
11. Ranjit N, Jones MK, Stenzel DJ et al. A survey of the intestinal transcriptomes of the hookworms, Necator americanus and Ancylostoma caninum, using tissues isolated by laser microdissection microscopy. Int J Parasitol 2006; 36:701-710.
12. Loukas A, Bethony J, Brooker S et al. Hookworm vaccines: past, present, and future. Lancet Infect Dis 2006; 6:733-741.
13. Kazazian HH, Jr. Mobile elements: drivers of genome evolution. Science 2004, 303:1626-1632.
14. Plasterk RH, Izsvak Z, Ivics Z. Resident aliens: the Tc1/mariner superfamily of transposable elements. Trends Genet 1999; 15:326-332.
15. Yang N, Zhang L, Kazazian HH, Jr. L1 retrotransposon-mediated stable gene silencing. Nucleic Acids Res 2005; 33:e57.
16. Brindley PJ, Laha T, McManus DP et al. Mobile genetic elements colonizing the genomes of metazoan parasites. Trends Parasitol 2003; 19:79-87.
17. Venter JC, Adams MD, Myers EW et al. The sequence of the human genome. Science 2001; 291:1304-1351.
18. Fischer C, Bouneau L, Coutanceau JP et al. Global heterochromatic colocalization of transposable elements with minisatellites in the compact genome of the pufferfish Tetraodon nigroviridis. Gene 2004; 336:175-183.
19. Laha T, Kewgrai N, Loukas A et al. The dingo nonlong terminal repeat retrotransposons from the genome of the hookworm, Ancylostoma caninum. Exp Parasitol 2006; 113:142-153.
20. Laha T, Loukas A, Wattanasatitarpa S et al. The bandit transposon, a mariner-like mobile genetic element from the genome of the hookworm, Ancylostoma caninum: inferred horizontal transmission between host and parasite. PLoS Negl Trop Dis 2007; 1(1):e35.
21. Kapitonov VV, Jurka J. RAG1 core and V(D)J recombination signal sequences were derived from Transib transposons. PLoS Biology 2005; 3(6):e181.
22. Malik H, Eickbush T. The RTE class of non-LTR retrotransposons is widely distributed in animals and is the origin of many SINEs. Mol Biol Evol 1998; 15(9):1123-34.

23. Laha T, Kewgrai N, Loukas A et al. Characterization of SR3 reveals abundance of non-LTR retrotransposons of the RTE clade in the genome of the human blood fluke, Schistosoma mansoni. BMC Genomics 2005; 6:154.
24. Youngman S, van Luenen HG, Plasterk RH. RTE-1, a retrotransposon-like element in Caenorhabditis elegans. FEBS Letters 1996; 380:1-7.
25. Drew AC, Minchella DJ, King LT et al. SR2 elements, nonlong terminal repeat retrotransposons of the RTE-1 lineage from the human blood fluke Schistosoma mansoni. Mol Biol Evol 1999; 16:1256-1269.
26. Malik HS, Burke WD, Eickbush TH. The age and evolution of non-LTR retrotransposable elements. Mol Biol Evol 1999; 16:793-805.
27. Dai L, Zimmerly S. The dispersal of five group II introns among natural populations of Escherichia coli. RNA 2002; 8:1294-1307.
28. Ueda K, Yamashita A, Ishikawa J et al. Genome sequence of Symbiobacterium thermophilum, an uncultivable bacterium that depends on microbial commensalisms. Nuc Acids Res 2004; 32:4937-4944.
29. Szak TS, Pickeral OK, Makalowski W et al. Molecular archeology of L1 insertions in the human genome. Genome Biol 2002; 3:10.
30. Plasterk RH, Izsvak Z, Ivics Z. Resident aliens: the Tc1/mariner superfamily of transposable elements. Trends Genet 1999; 15:326-332.
31. Lohe AR, De Aguiar D, Hartl DL. Mutations in the mariner transposase: the D,D(35)E consensus sequence is nonfunctional. Proc Natl Acad Sci USA 1997; 94:1293-1297.
32. Rosenzweig B, Liao LW, Hirsh D. Sequence of the C. elegans transposable element Tc1. Nucleic Acids Res 1983; 11:4201-4209.
33. Hoekstra R, Otsen M, Lenstra JA et al. Characterisation of a polymorphic Tc1-like transposable element of the parasitic nematode Haemonchus contortus. Mol Biochem Parasitol 1999; 102:157-166.
34. Robertson HM. The mariner transposable element is widespread in insects. Nature 1993; 362:241-245.
35. Witherspoon DJ, Robertson HM. Neutral evolution of ten types of mariner transposons in the genomes of Caenorhabditis elegans and Caenorhabditis briggsae. J Mol Evol 2003; 56:751-769.
36. Robertson HM, Walden KK. Bmmar6, a second mori subfamily mariner transposon from the silkworm moth Bombyx mori. Insect Mol Biol 2003; 12:167-171.
37. Halaimia-Toumi N, Casse N, Demattei MV et al. The GC-rich transposon Bytmar1 from the deep-sea hydrothermal crab, Bythograea thermydron, may encode three transposase isoforms from a single ORF. J Mol Evol 2004; 59:747-760.
38. Prasad MD, Nurminsky DL, Nagaraju J. Characterization and molecular phylogenetic analysis of mariner elements from wild and domesticated species of silkmoths. Mol Phylogen Evol 2002; 25:210-217.
39. Richardson JM, Dawson A, O'Hagan N et al. Mechanism of Mos1 transposition: insights from structural analysis. EMBO J 2006; 25:1324-1334.
40. Feschotte C, Pritham EJ. DNA transposons and the evolution of eukaryotic genomes. Annu Rev Genet 2007; 41:331-368.
41. Bao W, Jurka M, Kapitonov VV et al. New superfamilies of eukaryotic DNA transposons and their internal divisions. Mol Biol Evol 2009; (in press).
42. Chen S, Li X. Molecular characterization of the first intact Transib transposon from Helicoverpa zea. Gene 20008; 408:51-63.
43. Kapitonov VV, Jurka J. Molecular paleontology of transposable elements in the Drosophila melanogaster genome. Proc Natl Acad Sci U S A 2003; 100:6569-6574.
44. Stein LD, Bao Z, Blasiar D et al.The genome sequence of Caenorhabditis briggsae: A platform for comparative genomics. PLoS Biol 2003; 1(2):e45.
45. Ghedin E, Wang S, Spiro D et al. Draft genome of the filarial nematode parasite Brugia malayi. Science 2007; 317:1756-1760.
46. Zhou Y, Zheng H, Jin K et al. The genome of Schistosoma japonicum reveals unique features of host-parasite interface. Nature 2009; (in press).
47. Bessereau JL, Wright A, Williams DC et al. Mobilization of a Drosophila transposon in the Caenorhabditis elegans germ line. Nature 2001; 413:70-74.
48. Blaxter ML. Nematoda: genes, genomes and the evolution of parasitism. Adv Parasitol 2006; 54:101-195.
49. Dieterich C, Clifton SW, Schuster LN et al. The Pristionchus pacificus genome provides a unique perspective on nematode lifestyle and parasitism. Nat Genet 2008; 40(10):1193-1198.

Index

M

N

O

P

R

S

T

U